12 Real-Life Accounts of Transformation

Inspired by a True Story

Discover *the* Untold Potential *of* Your Own Story

JEFF BLOOMFIELD

Axon Publishing

Axon Publishing

Inspired by a True Story: Discover the Untold Potential of Your Own Story

For more information about this title or to order other books
and/or electronic media, contact the publisher:
Axon Publishing, LLC
info@braintrustgrowth.com

ISBN: 978-1-7337870-5-5 (Paperback)
978-1-7337870-6-2 (Hardback)
978-1-7337870-7-9 (eBook)

Contents

Introduction

Welcome to a journey unlike any other—a journey that takes you through the lives of remarkable, yet ordinary individuals who have faced seemingly insurmountable obstacles, and emerged not just as survivors, but as beacons of hope, resilience, and purpose. This book is about more than just reading stories. It's about connecting with your own story through the stories of others in a way that empowers you to see the beauty, the strength, and the extraordinary significance of your life, no matter where you've been or what you've faced.

We all live with a narrative running in the background of our minds. It's the sum of our experiences, choices, and the people we've encountered along the way. This narrative can lift us up or weigh us down, depending on how we choose to view it. But what if I told you that your narrative is far more profound than you've ever realized? What if the story of your life holds the key not only to your own growth, but to inspiring others? That's the heart of this book—the recognition that our stories, however ordinary they may seem on the surface, are woven with extraordinary threads of purpose and potential.

As an author, keynote speaker, and host of a podcast, I get the privilege of traveling the globe and meeting people that I might otherwise never cross paths with. And if there's one thing I've learned in all those conversations, it's this: everyone has a story, and every story carries the potential to change lives. But here's the catch—most of us are unaware of just how powerful our stories really are. We walk through life downplaying our experiences, dismissing the lessons we've learned, and ignoring the wisdom we've gained. We think we're just surviving, when in reality, we've been building a foundation of resilience and strength all along. Your story is no different.

What's waiting to be uncovered within your own narrative? As you turn these pages, you'll be introduced to people that I've had the privilege of interviewing on my show, *The Driving Change Podcast,* who have walked through fire and come out the other side, not unscathed, but transformed. You'll meet Chad, who lost his physical sight but found a vision more powerful than most of us could ever imagine. You'll walk with Dawn through the storms of life and discover the unshakable strength within her. You'll journey with Hilary as she battles the voices of self-doubt and the claws of addiction and emerges victorious in her own skin. Each of these individuals—and so many more—share their stories not as tales of defeat, but as triumphs of the human spirit.

But I'll let you in on a little secret; this book really isn't about *them.* It's about *you.* It's about the story you've lived, the challenges you've faced, and the purpose that lies within you, possibly still waiting to be uncovered. Too often, we think our stories are insignificant. We compare them to others, diminishing our own experiences and downplaying our own strength. But I'm here to tell you: ***your story matters***.

Let me explain why. Growing up on a farm in Ohio taught me lessons that were hard but invaluable. Life on the farm instilled in me perseverance, grit, and the meaning of pushing forward, even when it feels like the world is pressing you down. Every experience—no matter how small—became a thread in the larger tapestry of my life.

Each chore at 5:30 in the morning, each bale of hay I hoisted onto the rusty wagon in the blistering heat, each tear I shed when my Papaw passed from lung cancer and later, when my other grandfather died suddenly of a heart attack—all of these moments were pieces of a puzzle that, bit by bit, formed the mosaic of my life. A less-than-ideal family environment only added to this complexity, shaping and refining who I was becoming.

Then, years later, when I faced my own battle with cancer, I understood that these stories—our stories—are not just ours to keep. They're meant to be shared, to inspire, to spark hope in others who may be facing their own hardships. Every hardship, every triumph, every lesson has a purpose. It's the legacy we leave, the gift we give to the world.

Each one of us carries within us a unique blend of experiences, lessons, and wisdom that no one else in the world has. And while it may be easy to think that your struggles and victories are yours alone, the truth is that they are a gift to those around you. By sharing our stories, we invite others to see the world through our eyes, to learn from our mistakes, and to draw strength from our perseverance. Your story, in all its messiness and beauty, has the power to heal not only yourself but also those who hear it.

This book is a testament to that belief. It's a collection of stories that reflect the deepest parts of the human experience—the pain, the struggle, the joy, the triumph. But more than that, it's a guide for you to see your own life through a new lens. To recognize that every challenge you've faced, every mountain you've climbed, every valley you've walked through has been preparing you for something greater.

As you read, I encourage you to dive in with both feet. Engage with these stories, but more importantly, engage with your own. Reflect on the moments in your life that have shaped you. Consider how you've been molded by your experiences, how you've grown through your struggles, and how you've found strength in places you never thought possible. This book is not just a collection of stories—it's an invitation. An invitation to see your life in a new light, to embrace your story, and to step into your purpose with courage and conviction.

The people you'll meet in these pages are extraordinary, not because they are different from you, but because they've chosen to see their ordinary lives as significant. They've chosen to find purpose in their pain, to rise above their circumstances, and to use their stories to make a difference in the world. And that's something you can do too.

We live in a world where stories are constantly being told—through books, movies, social media, and conversations. But not all stories are created equal. Some stories inspire, while others discourage. Some lift us up, while others tear us down. The question is: which kind of story will you choose to tell? The people you'll encounter in this book have chosen to tell stories of triumph, resilience, and purpose, and by doing so, they have transformed not only their own lives but the lives of those around them.

So, as you embark on this journey, remember this: your story is powerful. It has the ability to inspire, to heal, to transform. It has the power to change your life and the lives of those around you. This book is a reminder that no matter where you've been or what you've faced, your story matters. And the world needs to hear it.

You see, stories have a way of reaching deep into our souls. They speak to us in ways that facts and figures never could. They remind us of our shared humanity, our common struggles, and our collective hope for a better tomorrow. Through the stories in this book, I hope you'll be reminded of the beauty in your own story. I hope you'll find the courage to embrace your experiences, both the good and the bad, and to use them as fuel for your future.

Thank you for joining me on this journey. Let's dive in together and discover the incredible power of your story.

CHAPTER 1

Andy Andrew's Story

From Homeless Under a Pier to Helping Others Find Purpose

In the tapestry of life, some threads seem ordinary, almost invisible, until they're woven into the grander design. Andy Andrews' story is one such thread—a seemingly ordinary life that became extraordinary through a series of tragic events, a mysterious mentor, and the discovery of timeless truths. It's a story that reminds us all of the power of perspective, the importance of the choices we make, and the belief that even in our darkest moments, there is a purpose waiting to be uncovered.

Andy Andrews is a name many recognize today—a prolific author, speaker, and storyteller whose words have touched millions. But before the accolades, before the bestsellers and the standing ovations, Andy's life was one of deep loss, profound struggle, and a transformative encounter that would change the course of his existence.

Growing up in Alabama, Andy's childhood was marked by frequent moves as his father, a minister, answered the call to serve in different communities. Every few years, the family would pack up and head to a new town, and while the scenery changed, the constant was his father's voice—a voice that carried not just sermons but stories. Stories that captivated young Andy, even when his mind wandered through the endless hours of church services. His father had a way of bringing the divine into the everyday, of turning the mundane into the miraculous, and it was this gift of storytelling that planted the first seeds in Andy's heart.

But like many seeds, it would take time—and a great deal of adversity—before they began to grow. Andy's early years were, by all accounts, typical. He wasn't the standout student, and he certainly didn't see himself as the best writer in his class. In fact, if someone had suggested that he would one day become a bestselling author, he likely would have laughed. Writing was something he did because he had to, not because he excelled at it. Yet, there was something about the way stories could cut through the noise and touch the heart that drew him in, even then.

Andy's life took a devastating turn when he was just 19. In a single, cruel year, he lost both of his parents—his mother to cancer, a slow and agonizing battle that drained the family of hope, and his father to a sudden car accident, a jarring shock that left Andy reeling. In the span of months, the foundation of his world crumbled, leaving him adrift in a sea of grief and confusion. The stability he had known was gone, and with it, the future he had once imagined.

Grief is a powerful force, and it can drive a person to places they never thought they would go. For Andy, it led to a downward spiral, one that saw him lose not just his parents, but his sense of self, his direction, and eventually, his home. In just two short years, Andy went from having a car and a trailer to living under a pier on the Gulf Coast, homeless and aimless, clinging to a threadbare existence. The once vibrant young man was now a shadow, haunted by the ghosts of his past and the uncertainty of his future.

It was during this dark night of the soul that Andy encountered a man who would change everything. His name was Jones, though no one called him Mister, and he was as much a mystery as he was a fixture in the small coastal town. People had seen him around for years, but no one knew much about him—where he came from, where he went, or how old he was. He carried a weathered suitcase and wore the same clothes every day, yet he seemed somehow above the grime of his surroundings, as if the dirt and despair of the world simply couldn't touch him.

One night, under the pier where Andy had made his makeshift home, Jones appeared. Andy had seen him before, like a shadow passing in the night, but he had never expected to meet him face to face, let alone have him take an interest in his life. Jones didn't offer Andy the kind of help most people do when they see someone down and out. He didn't give him a meal or a few dollars. Instead, he offered something far more valuable - truth. And not the kind of truth that soothes, but the kind that stings, the kind that gets under your skin and forces you to confront the parts of yourself you'd rather ignore.

Jones was the first person to look past Andy's circumstances and see the potential buried beneath the layers of pain and loss. He didn't sugarcoat his words; he told Andy things he didn't want to hear but desperately needed to. He spoke to Andy's heart, cutting through the self-pity and despair with a clarity that was both unsettling and invigorating. For the first time in a long while, Andy felt something stir within him—a glimmer of hope, a spark of purpose, though it was still far from clear what that purpose might be.

Jones also handed Andy three books—biographies of Winston Churchill, George Washington Carver, and Will Rogers. At first, Andy was skeptical. Biographies? They were the kind of books that had bored him to tears in school. But Jones insisted, telling him these weren't just biographies—they were adventure stories, mysteries, romances, and thrillers, all rolled into one. And more importantly, they were true.

Curiosity, and perhaps a sense of obligation, led Andy to open the first book. He didn't expect much, but as he read, something happened. The stories of these great men began to draw him in, not just because of their accomplishments, but because of their struggles. These were men who had faced incredible odds, who had been knocked down by life again and again but had somehow found the strength to get back up, to keep going, to make something of themselves.

As Andy read, he began to see patterns in their lives—choices they had made, principles they had lived by, often without even realizing it. Over time, Andy identified seven principles that seemed to be the common thread running through the lives of these extraordinary individuals. These principles weren't just habits or traits; they were foundational truths that guided their decisions and shaped their destinies.

For the first time since his parents' deaths, Andy began to feel a sense of purpose. He realized that his life didn't have to be defined by the tragedies he had endured, that he could choose a different path. It wasn't easy and climbing out from under that pier was a slow and painful process, both literally and metaphorically. But with each step, Andy began to rebuild his life, brick by brick, choice by choice, principle by principle.

He started small, taking odd jobs, renting a place, and eventually finding his way onto a stage. At first, it was as a comedian—because he knew that humor could open doors, could break down barriers. But as he gained confidence, he began to share more of what he had learned during those dark days, the principles that had pulled him back from the brink. It was a risky move—people came to laugh, not to be lectured—but the response was overwhelmingly positive. Audiences connected with the depth behind the jokes, the truth behind the humor.

One of those audience members was country music legend Kenny Rogers, who took Andy under his wing. Rogers was at the height of his career, filling arenas night after night, and he saw something in Andy that reminded him of his own journey—a raw talent, yes, but also a deep understanding of what really mattered in life. Rogers encouraged Andy to share more of his insights, to weave them into his performances, and even invited him to be his opening act, giving Andy a platform to refine his message.

This period of mentorship and growth led Andy to write *The Traveler's Gift*, a book that would go on to become a New York Times bestseller and change the lives of millions. The book tells the story of a man going through a personal crisis who is given the chance to travel through time, meeting with seven historical figures who each impart a principle that can change his life. It's a narrative steeped in history and wisdom, and it resonates with readers because it's rooted in the truth of Andy's own journey.

But Andy's story didn't end with *The Traveler's Gift*. He continued to write, his works including *The Butterfly Effect*, *The Noticer*, and the thought-provoking *How Do You Kill 11 Million People?*—each book a reflection of the principles he had discovered during those dark days under the pier. His ability to weave life lessons into engaging narratives has made him one of the most sought-after speakers and authors of our time.

Today, Andy Andrews is more than just a bestselling author. He's a thought leader, a mentor, and a noticer in his own right. His books, now translated into over 40 languages, have sold more than 20 million copies worldwide. His insights have been requested by U.S. presidents, and his teachings have reached audiences in every corner of the globe.

But no matter how far he's come, Andy has never forgotten those early lessons—lessons from a mysterious old man under a pier, from the biographies of great leaders, and from the trials and triumphs of his own life. He continues to share those lessons with the world, believing that everyone has the potential to turn their life around, to find their purpose, and to make a lasting impact.

Andy's story is a testament to the power of perspective, responsibility, and the choices we make. It's a story of hope, transformation, and the enduring truth that our past does not define our future. And it all began with a noticer, a suitcase full of books, and a young man who decided to read.

Reflection: Building a Legacy of Purpose and Influence

Andy Andrews' journey from homelessness to becoming a bestselling author and influential speaker is a powerful reminder of how our choices not only shape our lives but also create ripples that affect the lives of others. His story is an invitation to reflect deeply on how we can transform our challenges into opportunities to build a legacy that endures. Each step in Andy's journey offers profound insights into how we can lead lives of greater meaning and impact.

1. **Legacy and Long-Term Impact: Shaping the Future Beyond Yourself**

   When Andy was living under a pier, with nothing but the clothes on his back and the sorrow of losing his parents, the idea of leaving a lasting legacy seemed unimaginable. Yet, through small, deliberate choices—like reading the biographies of great leaders—Andy began to shape a life that would eventually touch millions. His story shows that no matter where we start, we have the power to create a legacy that reaches far beyond our immediate circumstances.

   - **Reflection:** Andy's transformation reminds us that legacy isn't built overnight—it's the result of consistent, purposeful actions. Each choice we make can either bring us closer to or further from the impact we wish to leave behind. This perspective encourages us to consider the long-term consequences of our daily decisions. Are you investing in things that will outlast you, or are you consumed by the immediate and the fleeting?

   - **Strategy:** Begin by identifying a cause, mission, or value that resonates deeply with you. This could be something related to your work, your community, or your family. Reflect on how your current actions align with this mission. Are you building something that will endure? Start small by setting aside time each day to work on a project that contributes to your legacy, whether it's mentoring someone, writing, or creating something that will benefit others in the long run. Regularly review your progress and adjust your actions to ensure they align with your long-term vision.

2. **Taking Ownership of One's Narrative: Rewriting Your Story**

   After losing both of his parents and becoming homeless, Andy could have easily succumbed to despair, allowing his circumstances to define him. Instead, he chose to take ownership of his narrative. By confronting his reality and making intentional choices to change his path, Andy redefined his life story. This act of taking control is a powerful reminder that we are not merely products of our circumstances—we are the authors of our own lives.

- **Reflection:** We often forget that we have the power to rewrite our stories. It's easy to feel trapped by past decisions, failures, or external circumstances. But Andy's story teaches us that it's never too late to take control. Reflect on areas in your life where you feel stuck or defined by past events. What would happen if you chose to rewrite those narratives? How would your life change if you viewed your challenges as the starting point of a new chapter rather than the end of the story?

- **Strategy:** Take time to write down the current narrative you hold about your life. Identify the elements that feel disempowering or limiting. Then, rewrite this story with you as the hero—focusing on your strengths, the lessons you've learned, and the opportunities that lie ahead. Revisit this new narrative regularly, especially in moments of doubt, and let it guide your decisions and actions. Over time, you'll find that this shift in perspective empowers you to take bold steps towards the life you desire.

3. **Mentorship and the Power of Guiding Others: Passing on the Wisdom**

Throughout his life, Andy was influenced by mentors who saw potential in him even when he couldn't see it in himself. These relationships played a crucial role in his transformation, and they highlight the profound impact that mentorship can have on our lives. But Andy's story doesn't stop there—he also became a mentor, passing on the wisdom he gained to others. This cycle of receiving and giving guidance is essential for personal growth and for building a legacy that lasts.

- **Reflection:** Mentorship is not just about receiving guidance; it's also about becoming a source of wisdom and support for others. Andy's story encourages us to reflect on the mentors who have shaped our lives and consider how we can pay that forward. Who are the people in your life who have helped you grow? And how can you extend that same support to someone else? Mentorship is a two-way street, and by engaging in this practice, we not only grow ourselves but also contribute to the growth of others.

- **Strategy:** Identify someone in your life who could benefit from your experience and wisdom. This could be a colleague, a friend, or even a family member. Make a conscious effort to offer them guidance, whether it's through formal mentoring sessions or simply being available for advice and support. At the same time, seek out mentors who can help you navigate your current challenges. Building these relationships will not only enrich your life but also create a ripple effect, extending your legacy through the lives you touch.

4. **Wisdom Through Learning from the Past: Drawing Lessons from History**

Andy's transformation was sparked by reading the biographies of great leaders who had faced incredible challenges. These stories provided him with the principles and values that guided his journey. By learning from the past, Andy was able to navigate his own challenges with greater wisdom and perspective. His story reminds us that the past holds valuable lessons that can inform our present and future decisions.

- **Reflection:** The past is a rich source of wisdom, offering lessons that can guide us through our own challenges. Andy's story illustrates the power of learning from those who have gone before us. Reflect on the people, historical figures, or even family members whose stories resonate with you. What can you learn from their experiences? How can their challenges and triumphs inform your own decisions?

- **Strategy:** Commit to regularly studying the lives of individuals who have faced similar challenges or who embody the values you aspire to. This could involve reading biographies, watching documentaries, or engaging in conversations with those more experienced. Extract key principles from their stories and apply them to your life. Create a personal guidebook of wisdom that you can refer to when facing difficult decisions or uncertain times. By anchoring yourself in the lessons of the past, you equip yourself to make wiser choices in the present.

5. **Clarity of Vision: Seeing Beyond the Immediate**

Andy's success was deeply rooted in his ability to see beyond his immediate circumstances. Even when he was at his lowest point, he maintained a vision for his life that extended far beyond his current reality. This clarity of vision was crucial in guiding his decisions and actions, allowing him to stay focused on his long-term goals despite the challenges he faced.

- **Reflection:** Having a clear vision for your life is essential for navigating challenges and staying true to your goals. Andy's story reminds us that when we have a strong sense of where we're headed, we're better equipped to make decisions that align with our long-term aspirations. Reflect on your own vision for your life. Is it clear and compelling, or does it feel vague and uncertain? How might your decisions change if you had a clearer sense of direction?

- **Strategy:** Take time to articulate a clear vision for your life. Where do you want to be in 5, 10, or even 20 years? What do you want to accomplish, and what values do you want to uphold? Write down this vision and break it into actionable steps that you can start taking today. Regularly revisit and refine your vision as you grow and as circumstances change. By maintaining clarity of vision, you ensure that your daily actions contribute to your larger life goals, allowing you to navigate challenges with confidence and purpose.

To learn more about Andy's work and his journey:

https://drivingchangepodcast.com/podcast/andy-andrews-finding-and-choosing-hope/

https://andyandrews.com/

https://wisdomharbour.com/

http://creatingmeasurableresults.com/

CHAPTER 2

Chad Foster's Journey

From Darkness to Light

In the heart of Tennessee, where the mountains cradle the sky and the air is thick with the scent of pine and possibility, a young boy named Chad Foster grew up with the world seemingly at his feet. From the outside, his childhood appeared as any other; days filled with laughter, scraped knees, and the endless curiosity that only the boundless imagination of youth can bring. But there was something else, something lurking in the shadows that only he and his family were beginning to understand.

From the tender age of three, Chad's parents noticed something peculiar. In dimly lit rooms, Chad would stumble, his tiny feet betraying him as he tripped over toys and bumped into furniture. Nights brought no respite; the dark seemed to engulf him, swallowing the familiar shapes and leaving him blind to the world around him. Concerned, his parents took him to Duke University, where doctors delivered a diagnosis that would alter the trajectory of his life—retinitis pigmentosa, a degenerative eye condition that would slowly steal his sight.

Chad was the first in his family to bear this burden, the odds were a cruel twist of fate. The doctors, in their clinical detachment, advised his parents to place him in a special school for the blind, a suggestion that might have consigned Chad to a life of limitations. But his parents, with the fierce love and determination that only parents can muster, refused to let their son be defined by his diagnosis. Instead, they enrolled him in soccer, teaching him that the world may take away his sight, but it could never steal his spirit.

His father, a man of old-school values and tough love, pushed Chad beyond what seemed reasonable, not out of cruelty, but out of necessity. He knew that life would demand more from his son, and he was determined to ensure Chad would be ready. And so, Chad's childhood became a series of trials—both physical and emotional—as he learned to navigate a world that was slowly disappearing from view.

As he grew older, the reality of his condition became harder to ignore. In high school, Chad faced the social sting of not being able to drive at night, of missing out on the camaraderie of football games under the Friday night lights, and the simple joy of playing basketball in a well-lit gym. These were the years when life is supposed to be about discovery and possibility, but for Chad, it was about coming to terms with what was being taken away. The frustration and anger began to build, simmering just beneath the surface.

The real breaking point came in college. Chad had dreams of entering the medical field, of helping others as he had been helped, but those dreams came crashing down when he lost the last remnants of his sight. At 21, he found himself plunged into darkness—not just the physical kind, but the kind that seeps into the soul, leaving a person questioning their worth, their future, and the very meaning of their existence. His world, once full of possibilities, was now a tight, suffocating space filled with despair.

For a time, Chad lost himself in that darkness. The vision of a bright future was replaced with a cloud of uncertainty. He partied too much, cared too little, and found himself on the brink of abandoning his education altogether. Why bother, he thought, when the life he had envisioned was now out of reach? The anger and frustration bubbled over, manifesting in ways that were both self-destructive and counterproductive.

But even in the depths of his despair, there were glimmers of light—moments that would become the catalysts for his transformation. One such moment came during a visit from his cousin Mark, a man Chad had always looked up to. His cousin, in a moment of tough love, told him bluntly that if he didn't change his ways, he would become a loser, just like the company he was keeping. The words hit hard, piercing through the fog of self-pity that had surrounded Chad for so long.

Yet, the true turning point came when Chad, at 23, decided to get his first guide dog. He traveled to Michigan to attend the Leader Dogs for the Blind program, still clinging to his victim mentality. But what he encountered there shattered the walls of his self-imposed prison. Among the other participants were individuals who faced challenges far greater than his own—those who were not only blind but also deaf, mentally impaired, or battling chronic illnesses. These individuals, who had every reason to give up, were there to gain independence, to reclaim their lives.

The experience was nothing short of transformative. Chad saw in these people a strength and courage that dwarfed his own. They were living testaments to the power of the human spirit, and their example forced Chad to reevaluate everything he had been telling himself. He began to shift his focus from what he had lost to what he still had—a healthy body, a sharp mind, and most importantly, a choice. A choice to either remain in the darkness or to step into the light of possibility.

It was here that Chad began to understand a profound truth: happiness is not a feeling bestowed upon us by circumstance, but a choice we make every day. It is born not of abundance, but of gratitude. And so, Chad chose to be grateful—not just for what he had, but for the challenges that had shaped him. He learned that the narrative he had been telling himself was flawed, and it was up to him to rewrite it.

Armed with this new perspective, Chad returned to college with a renewed sense of purpose. He had to relearn how to learn—no longer able to rely on his sight, he developed new strategies, adapting and overcoming. His mother, embodying the very essence of selfless love, read every single one of his college textbooks onto cassette tapes, her voice guiding him through the academic challenges that once seemed insurmountable. With her support and his newfound determination, Chad excelled, earning straight A's and a spot on the Dean's List.

His success in college led to opportunities in the corporate world that might have seemed impossible just a few years earlier. He landed a job at a top consulting firm in Atlanta, a city that was as unfamiliar and daunting to him as his new life as a blind man. But Chad faced these challenges head-on, learning the bus and train systems, figuring out how to navigate a new world with his guide dog by his side, and proving to himself and others that he could not only survive but thrive.

As Chad's career advanced, so did his understanding of his own potential. He excelled in roles that required not just technical skill but also leadership and vision—qualities that had been honed through years of overcoming adversity. His accomplishments in the corporate world culminated in an opportunity to attend Harvard Business School's Leadership Program, where he became the first blind person to graduate. It was here, under the guidance of esteemed leaders like Bill George, that Chad truly began to understand the power of his story.

At Harvard, Chad learned that true leadership is born not from titles or achievements but from authenticity and purpose. It is about using one's life experiences—especially the painful ones—to inspire and uplift others. And for Chad, this meant embracing his blindness not as a curse but as a gift, one that had been given to him for a reason. He realized that his journey was not just about overcoming obstacles for his own sake, but about showing others that they too could turn their trials into triumphs.

This epiphany led Chad to a new chapter in his life, one where he would step onto stages around the world to share his story. His message was clear and powerful: we are not defined by the challenges we face, but by how we choose to respond to them. In his keynote speeches and in his book, "Blind Ambition," Chad offered a roadmap for others to follow—one that was marked by resilience, determination, and the unwavering belief that within each of us lies the power to transform our greatest challenges into our greatest strengths.

Chad's story is not just about blindness; it is about vision—the kind that sees beyond the immediate, beyond the physical, to the possibilities that lie within each of us. His life is a testament to the idea that no matter how dark the path may seem, there is always a way forward if we are willing to look within and harness the power of our own stories.

As Chad continues to inspire audiences around the globe, his message remains as clear as ever: the obstacles in our lives are not there to stop us but to shape us into the people we were always meant to be. And for Chad, the darkness that once seemed so overwhelming has become the very light that guides his journey—a journey that continues to touch the lives of countless others, encouraging them to see beyond their limitations and to embrace the boundless possibilities that await.

Reflection: Transforming Adversity into Opportunity and Connection

Chad's journey is a powerful example of how we can transform adversity into an opportunity for growth and connection. His story teaches us that our greatest challenges can become the very thing that propels us forward, allowing us to discover strengths we never knew we had. Chad's experience also highlights the importance of community and the narratives we craft about our lives. By reframing his blindness and embracing the support of others, Chad turned his life's greatest obstacle into his greatest opportunity.

1. **Reframing Adversity as Opportunity: The Catalyst for Growth**

   When Chad was diagnosed with retinitis pigmentosa, a degenerative eye condition that would eventually leave him blind, it would have been easy to see this as the end of his dreams. However, Chad chose to view his blindness as a unique challenge—one that he was determined to overcome. By reframing his adversity, Chad unlocked new possibilities for growth and achievement, ultimately leading him to a successful career and a life of purpose.

   - **Reflection**: Chad's ability to see his blindness not as a limitation but as an opportunity is a profound reminder of the power of perspective. We all face challenges that can feel insurmountable, but Chad's story shows us that by changing the way we view our obstacles, we can transform them into catalysts for growth. Reflect on a challenge you're currently facing. How are you viewing this situation? Is it a barrier, or could it be a steppingstone to something greater?

- **Strategy**: The next time you encounter a challenge, consciously shift your perspective. Instead of focusing on what's being taken away, ask yourself, "What new opportunities does this challenge present? How can I grow from this experience?" Write down your thoughts and identify specific actions you can take to turn the challenge into an opportunity. This practice will not only shift your mindset but also empower you to take proactive steps toward overcoming your obstacles and achieving your goals.

2. **Embracing the Unknown: Finding Growth in Discomfort**

Throughout his journey, Chad faced numerous situations where he had to step into the unknown—whether it was navigating a new city as a blind person or speaking in front of an audience at Harvard. Each time, Chad discovered new strengths and capabilities he didn't know he had. His story is a powerful testament to the idea that real growth often lies on the other side of discomfort.

- **Reflection**: Stepping into the unknown can be terrifying, but it's often where we find the greatest opportunities for growth. Chad's story encourages us to embrace discomfort and uncertainty as necessary parts of our journey. Reflect on areas in your life where you've been avoiding discomfort. What might you gain if you were willing to step into the unknown and face these challenges head-on?

- **Strategy**: Identify one area in your life where you've been avoiding discomfort, whether it's a new challenge at work, learning a new skill, or having a difficult conversation. Make a deliberate decision to confront this discomfort head-on. Approach it with the mindset that discomfort is not something to be feared but something to be embraced as a sign of growth. Reflect on the outcomes and continue to seek out situations that push you beyond your comfort zone. Over time, you'll build resilience and discover strengths you never knew you had.

3. **Building Resilient Networks: The Power of Connection**

Chad's journey was supported by a network of individuals who provided him with the strength and encouragement he needed to succeed. From his mother's unwavering support to his cousin's tough love, these relationships were crucial in helping Chad navigate the challenges of his blindness. His story highlights the importance of building and nurturing resilient networks that can support us through life's difficulties.

- **Reflection**: None of us achieve success alone. Chad's story reminds us that the people we surround ourselves with play a crucial role in our ability to overcome challenges and achieve our goals. Reflect on the relationships in your life. Are they providing you with the support and encouragement you need, or are there gaps in your network that need to be filled?

- **Strategy**: Take time to evaluate the strength of your current support network. Identify the individuals who uplift and encourage you and make a conscious effort to nurture those relationships. If you find gaps in your network, seek out new connections by joining groups, attending events, or reaching out to potential mentors. Strengthen your ties by offering support in return, creating a resilient network that is mutually beneficial. By building and maintaining strong connections, you ensure that you have the support you need to navigate life's challenges.

4. **Practicing Daily Joy: Cultivating Gratitude in the Present Moment**

One of the most significant shifts in Chad's life occurred when he began to focus on what he still had rather than what he had lost. By practicing daily joy and gratitude, Chad was able to find contentment and happiness, even in the face of significant challenges. His story teaches us that joy is not something that happens to us—it's something we actively cultivate.

- **Reflection**: Chad's practice of daily joy and gratitude is a powerful reminder that our happiness is often within our control, regardless of our circumstances. Reflect on your own life. How often do you focus on what's going right versus what's going wrong? How might your perspective shift if you made a conscious effort to practice gratitude and find joy in the present moment?

- **Strategy**: Develop a daily practice of recognizing and celebrating the small joys in your life. Each morning or evening, write down three things you are grateful for. Focus on the present moment and find joy in simple pleasures, whether it's a conversation with a friend, a moment of quiet, or a small accomplishment. Over time, this practice will help you maintain a positive outlook and build resilience against life's challenges, allowing you to experience greater contentment and happiness.

5. **Crafting Empowering Self-Talk: Shaping Your Reality through Narrative**

Chad's transformation was deeply connected to the narratives he created about his life. By choosing to see himself as a visionary rather than a victim, Chad was able to rewrite his story in a way that empowered him to take control of his future. His experience highlights the power of self-talk in shaping our reality and the importance of crafting narratives that serve our growth.

- **Reflection**: The stories we tell ourselves about our lives have a profound impact on how we experience the world. Chad's story reminds us that by choosing empowering narratives, we can transform our reality and achieve our goals. Reflect on the current narrative you hold about your life. Is it empowering, or does it hold you back? How might your life change if you chose to rewrite this narrative in a way that emphasizes your strengths and potential?

- **Strategy**: Take time to write down the narrative you currently hold about your life. Identify the elements that feel disempowering or limiting. Then, rewrite this story with you as the hero—focusing on your strengths, the lessons you've learned, and the opportunities that lie ahead. Practice affirming this new narrative daily, especially in moments of doubt or challenge. Over time, you'll find that this shift in perspective empowers you to take bold steps toward the life you desire, shaping your reality in a way that aligns with your goals and aspirations.

As you reflect on Chad Foster's journey, consider how these lessons might apply to your own life. Each challenge you face, each obstacle in your path, is an opportunity to write a new chapter in your story. The question is, what story will you choose to tell? By shifting your perspective, embracing discomfort, nurturing your community, practicing gratitude, and crafting an empowering narrative, you can navigate life's challenges with clarity, purpose, and resilience.

To hear Chad tell his story and learn more about his journey:

> https://drivingchangepodcast.com/podcast/overcoming-adversity-and-finding-purpose-chad-foster/
>
> https://chadefoster.com/
>
> https://chadefoster.com/book/blind-ambition/

CHAPTER 3

Dawn Mussallem's Story

A Journey of Heart, Spirit, and Unyielding Resilience

In the quiet, picturesque town of Johnstown, Ohio, a little girl named Dawn Mussallem had dreams that reached far beyond the horizons of her small community. Unlike most children who indulged in fleeting fantasies, Dawn's aspirations were rooted in a deep commitment to longevity and wellness. Even at the tender age of five, she had a singular vision: to one day see her name on a Smucker's jar, a symbol reserved for those who lived to be 100 years old. This wasn't just a childish whim; it reflected a profound inner drive that would shape the rest of her life.

Picture Dawn, a determined young girl, meticulously packing the healthiest lunches for her school trips while her classmates reached for cookies and chips. She gravitated toward spirulina bars, Alta Dena yogurt, and the oats section of the local health food store. Dawn wasn't merely interested in being healthy; she was laying the foundation for a life dedicated to wellness and longevity. This commitment was not just about living long but living well—an ethos that would carry her through the darkest days of her life.

As Dawn grew older, her passion for health evolved into a deep interest in medicine. She pursued studies in nutrition and exercise physiology in college, driven by the belief that the body was a temple deserving of the utmost care. After college, she landed what she considered her dream job at the Cooper Institute, where she studied centenarians—people who had lived to be 100 years old or more and were still running long distances. It was a job that seemed tailor-made for her, combining her love for wellness with her desire to understand the secrets of longevity.

But Dawn's journey wasn't destined to be straightforward. After a few years, she found herself drawn to naturopathic medicine and enrolled in a naturopathic school in Arizona. She loved her training, but she knew that her true calling required a deeper understanding of conventional medicine. She wanted to bridge the gap between traditional and alternative approaches, so she enrolled in Midwestern University in Arizona to pursue a traditional medical degree.

Just weeks into her studies at Midwestern, life took a harrowing turn. Dawn, a woman who regularly ran 10 to 14 miles a day and climbed Camelback Mountain twice daily, began to feel weak, tired, and breathless. She was coughing and struggling to keep up with her rigorous exercise routine. Concerned, she visited a doctor, but her symptoms were dismissed as stress-related, likely brought on by the demands of medical school. Another doctor echoed the same sentiment, attributing her condition to the power of suggestion—after all, she was studying pulmonary health at the time.

But Dawn knew her body, and she knew something was terribly wrong. Three more weeks passed, and her condition worsened. Then, one day in November 2000, as she climbed the stairs to her apartment, her body gave out, and she collapsed. Dawn was rushed to the emergency room, where a battery of tests revealed the devastating truth: she had stage four diffuse large B-cell non-Hodgkin's lymphoma. A 16-centimeter tumor was crushing her left lung, shifting her heart and other lung to the opposite side of her chest, and wrapping itself around her heart, preventing it from beating properly.

Imagine the terror of that moment—a vibrant, 26-year-old woman at the pinnacle of her dreams, suddenly faced with an almost certain death sentence. The first doctor who saw her was blunt and pessimistic. He told Dawn that her cancer was very advanced and difficult to treat, especially given how far it had progressed. When a friend—who would later become her husband—asked how long she had, the doctor coldly replied, "Three months."

Three months. For the girl who dreamed of living to be 100, this was an unfathomable prognosis. But Dawn wasn't ready to accept it. She fired that doctor immediately, determined to find someone who believed in her as much as she believed in herself. The next day, Dr. Paul, her new doctor, entered her room. He was different. He asked Dawn if she was willing to fight, and when she said yes, he promised to be by her side every step of the way. He pointed to a picture of Jesus by her bed and said, "He's really the only one who knows the answer here, but we're going to fight this."

With that, Dawn's grueling battle began. She underwent emergency surgery to remove part of the tumor from her heart and lung, followed by four months of intensive chemotherapy, high-dose chemotherapy, a bone marrow transplant, and two months of radiation therapy. Remarkably, she didn't quit medical school. Her college was supportive, adjusting her schedule so she could take exams during her treatment. Despite the intense and debilitating treatments, Dawn never felt sick. Instead, she experienced a heightened sense of existence, a direct connection to humanity that left her feeling more alive than ever.

During her bone marrow transplant, which at the time was a much different and more isolating experience than it is today, Dawn was confined to her room for weeks. She was essentially the "girl in the bubble," unable to leave her room except for brief, early morning escapes to ride a stationary bike that the oncology team had brought in for her. They even placed a bike at the nurse's station, so she could look out at the vista of the mountains while she exercised. Through it all, Dawn maintained her studies, taking tests in her room while receiving chemotherapy.

Dr. Dan Rubin, a naturopathic doctor she had met in Arizona, became part of her treatment plan, integrating naturopathic therapies with her conventional treatment. Together, they ensured there were no harmful interactions between the therapies, and Dawn continued to feel surprisingly well.

Despite the dire predictions, Dawn defied the odds. By 2001, she had completed her radiation treatments and was back to doing research and continuing her medical studies. She even got engaged to the man who had been by her side throughout her battle. The doctors had warned her that pregnancy might not be possible, but once again, Dawn proved them wrong. She became pregnant and, to her astonishment, didn't realize it until she was five months along because she hadn't thought it was possible.

But the joy of pregnancy was tempered by a new challenge. When it came time to deliver her daughter, Dawn found the experience far more difficult than she had anticipated. She had always been fit, running, and exercising throughout her pregnancy just as she had during her cancer treatments, but the delivery was unusually hard. Afterward, she was so weak that she couldn't even bathe her newborn daughter. She felt eerily similar to how she had felt before her cancer diagnosis, and her worst fears were confirmed when she nearly passed out in her kitchen just weeks after giving birth.

Rushed to the emergency room once again, Dawn was diagnosed with severe heart failure. Her heart was functioning at just 8%, a level so low that it was barely compatible with life. If she hadn't been so physically fit, she likely would have experienced symptoms much earlier, but now it was clear that her heart couldn't sustain her for long.

The doctors delivered another grim prognosis: she wouldn't survive more than a year with her failing heart. Devastated, but not defeated, Dawn sought out the best care she could find at the Mayo Clinic, where doctors gave her hope. They started her on medications that improved her functional capacity and quality of life. Slowly, her heart began to recover. By 2004, Dawn was back in medical school, finishing her training, and eventually starting a residency at the Mayo Clinic.

Residency was a struggle. Dawn didn't have the stamina of her colleagues, but she pushed through, driven by her desire to help others. By 2006, her heart failure had worsened, and she was listed for a heart transplant. Doctors implanted a device to help her heart beat more effectively, but in 2008, Dawn faced her greatest loss yet—her husband died suddenly of cardiac arrest.

The loss was crushing. It was the hardest thing Dawn had ever been through, and it took a toll on her already fragile heart. Yet, in a twist of fate, her heart began to improve in the years following his death. The device that had been implanted started to help her heart remodel, and her ejection fraction—the measure of how well her heart was pumping—began to climb from 8% to 18%, then 20%, and eventually to 30–40%. Her heart was recovering, and she was taken off the transplant list.

Dawn threw herself into her work, becoming a hospitalist at Mayo Clinic. She thrived on the adrenaline and the opportunity to save lives, touching the hearts of her patients with her high energy and deep compassion. But in 2015, her health took another sharp decline. While driving with her daughter in Cleveland, Dawn felt dizzy and unwell. Tests confirmed that her heart was failing again, and she began experiencing more frequent complications.

One day, while preparing to give a presentation to her colleagues at Mayo Clinic, Dawn's legs began to shake uncontrollably. She was too weak to stand, so she was relieved when the room was set up for everyone to sit. But as she began her presentation, she found herself unable to coordinate her movements. The room started to go fuzzy, and then everything went black. Dawn had gone into complete ventricular fibrillation, a type of cardiac arrest. She was shocked back to life by her defibrillator, which went off multiple times before finally reviving her.

Despite the harrowing experience, Dawn's faith and determination never wavered. But her heart continued to fail, leading to one complication after another. By July 2019, after a particularly serious procedure that caused her to lose vision in her left eye, it became clear that she needed a heart transplant. She was listed for a transplant in December 2019 and was told to prepare for a long wait.

The wait was excruciating. Dawn's condition worsened, and she contracted COVID-19, which left her even more vulnerable. By January 2021, she was too sick to stay at home and was admitted to the hospital for IV medications to help her heart pump. Then, on February 4th, 2021, her doctor came into her room with the news she had been waiting for: a donor heart had been found.

But the news came with a caveat—the donor was considered high-risk due to hepatitis C. Dawn was faced with an agonizing decision. After praying for guidance, she decided to accept the heart. As she was wheeled into the operating room, she made eye contact with her surgeon, Dr. Sareyyupoglu, and felt an overwhelming sense of peace. There was no fear, only confidence that she was making the right choice.

When Dawn awoke after the surgery, she experienced what she could only describe as a spiritual awakening. Her new heart was beating strongly, and every cell in her body seemed to be vibrating with a new, heightened energy. She named her new heart "Grace," a name that came to her in a dream where she saw a vision of pure bliss and innocence. The dream was accompanied by the sound of the song "Grace" playing on her instrumental music playlist, and when she checked her email that night, she found a message with the subject line "Full of Grace."

In the days following her transplant, Dawn experienced a brief period of anger and frustration, likely due to the high doses of steroids she was on. But the dream of Grace brought her peace and acceptance, and she embraced her new heart with gratitude and love.

Dawn's recovery was nothing short of miraculous. Determined to push the limits of what was possible, she set a goal to run a marathon exactly one year after her transplant. Her doctors and colleagues were skeptical, but Dawn was resolute. She began training, running a 5K just four months after her transplant, climbing Camelback Mountain at four and a half months, running a 10-mile race at eight months, and completing a half marathon at 10 months.

On her one-year anniversary, Dawn crossed the finish line of a full marathon. As she did, she noticed a construction sign at the finish line with the word "Grace" written on it. Her race number was 365, symbolizing the 365 days she had lived with her new heart. Dawn's marathon was more than just a personal achievement; it was a tribute to her organ donor, a celebration of life, and a testament to the power of the human spirit.

Dawn Mussallem's journey is a story of unparalleled resilience, faith, and determination. She has faced unimaginable challenges—cancer, heart failure, the loss of her husband—and has emerged stronger each time. Her life is a testament to the power of purpose, love, and the unbreakable human spirit. Through it all, Dawn has remained focused on helping others, using her experiences to inspire and motivate her patients and everyone she meets.

Her story teaches us that no matter how dire the circumstances, we have the power to overcome them. With faith, determination, and love, we can face any challenge and come out the other side stronger and more determined to live a life full of purpose and grace. Dawn's journey is far from over, and as she continues to inspire others, there is no doubt that she will leave an indelible mark on the world, one heartbeat at a time.

Dawn didn't just face her challenges—she transformed them into opportunities to live a life of profound impact and meaning. Her story isn't just about surviving—it's about thriving, even in the face of seemingly insurmountable odds. Dawn's journey invites us to consider how we, too, can navigate our own lives with the same grace, strength, and purpose.

Reflection: Living with Purpose, Resilience, and Grace

Dawn's life is a story of unparalleled resilience, unwavering faith, and a deep commitment to living with purpose. Her journey through life-threatening illness, loss, and the challenges of heart failure is a testament to the power of purpose, resilience, and gratitude. Dawn's story inspires us to embrace our challenges as opportunities for growth and to live each day with intentionality and grace.

1. **Embrace Purpose in Every Challenge: Turning Obstacles into Steppingstones**

   When Dawn was diagnosed with stage four diffuse large B-cell non-Hodgkin's lymphoma, she faced what seemed like an insurmountable challenge. Instead of succumbing to despair, Dawn chose to embrace her purpose even more fiercely, using her battle with cancer as a steppingstone to help others. Her story teaches us that purpose can be found even in our darkest moments, and it's what gives meaning to our struggles.

- **Reflection:** Dawn's journey reminds us that purpose is what transforms obstacles into opportunities. Reflect on the challenges you've faced in your own life. How did they shape your sense of purpose? What might change if you viewed your current challenges as opportunities to deepen your commitment to your purpose? Purpose is not just about grand achievements; it's about finding meaning in the everyday struggles and using them to fuel your growth and impact.
- **Strategy:** Identify a challenge you're currently facing and ask yourself how it aligns with your deeper purpose. How can you use this challenge to grow and contribute to others? Write down the ways in which this obstacle might be a steppingstone toward something greater. Reconnect with your purpose daily, reminding yourself that it is what gives meaning to both the highs and lows of your journey. Use this connection to guide your actions and decisions, turning challenges into catalysts for growth.

2. **Cultivate a Mindset of Resilience: Choosing Strength in Adversity**

Dawn's resilience was tested time and again, from her battle with cancer to the challenges of heart failure. Despite the odds, she chose to fight, driven by the belief that she could beat the odds and continue to live a life of purpose. Her story reminds us that resilience is not just about surviving adversity—it's about choosing strength, even when the situation seems hopeless.

- **Reflection:** Resilience is often seen as a trait, but Dawn's story shows us that it's also a choice—a decision to face adversity with strength and determination. Reflect on how you respond to adversity. Do you allow it to dictate your mindset, or do you choose resilience like Dawn? What would it look like to approach your current challenges with a mindset of strength and perseverance?

- **Strategy:** When faced with adversity, make a conscious choice to approach it with resilience. This means acknowledging the difficulty of the situation but refusing to let it define you. Create a resilience toolkit that includes practices like positive self-talk, seeking support from others, and focusing on small, manageable steps forward. Regularly engage in activities that build your mental and emotional strength, such as meditation, exercise, or journaling. By choosing resilience, you empower yourself to face any challenge with courage and determination.

3. **Find Strength in Vulnerability: Embracing Support and Connection**

Throughout her journey, Dawn leaned on her faith, family, and friends for support. She understood that vulnerability was not a weakness, but a source of strength. By embracing her vulnerability, Dawn was able to connect more deeply with those around her and draw on their support to face her challenges. Her story teaches us that true strength often comes from our willingness to be vulnerable and seek help when we need it.

- **Reflection:** Dawn's story challenges the notion that strength means facing challenges alone. Instead, it shows us that vulnerability—allowing others to support us—is a key component of resilience. Reflect on the areas of your life where you might be resisting vulnerability. How might your life change if you allowed yourself to be more open to support and connection?

- **Strategy:** Practice embracing vulnerability in your daily life. Start by identifying areas where you feel isolated or burdened and consider how opening up to others might bring relief and strength. Reach out to trusted friends, family, or professionals for support. Share your struggles and allow others to help you. By doing so, you not only lighten your load but also deepen your connections with those around you. Remember that vulnerability is a sign of courage, not weakness, and it's often through this openness that we find our greatest strength.

4. **Practice Gratitude in Every Circumstance: Finding Joy in the Journey**

Even in the midst of her most difficult battles, Dawn made a conscious effort to practice gratitude. She focused on the blessings in her life, no matter how small, and found joy in the simple moments. Her ability to maintain a grateful heart, even in the face of adversity, was a key factor in her resilience and overall well-being. Dawn's story teaches us that gratitude is a powerful tool for cultivating joy, no matter our circumstances.

- **Reflection:** Dawn's practice of gratitude, even during the darkest times, is a reminder that joy is a choice we make each day. Reflect on how often you focus on the positives in your life versus the challenges. How might your perspective shift if you made gratitude a daily practice? Gratitude doesn't ignore the difficulties—it simply chooses to focus on the blessings within them.

- **Strategy:** Develop a daily gratitude practice that helps you find joy in every circumstance. Each morning or evening, write down three things you are grateful for, no matter how small. Make an effort to express gratitude to others, whether through a note, a conversation, or a simple acknowledgment. Over time, you'll find that this practice shifts your focus from what's lacking to what's abundant, allowing you to experience greater joy and contentment, even in challenging times.

5. **Embrace Lifelong Growth: Continuing to Learn and Evolve**

   Despite the immense challenges she faced, Dawn never stopped growing. She continued to seek out new opportunities for learning and self-improvement, whether it was through her work as a physician or her personal spiritual journey. Her story reminds us that growth is a lifelong process, and it's through this continuous learning that we find fulfillment and purpose.

   - **Reflection:** Dawn's commitment to lifelong growth, even in the face of significant challenges, is a powerful reminder that we are never done growing. Reflect on your own journey. Are you actively seeking opportunities for growth, or have you become complacent? How might your life change if you approached each day as a chance to learn and evolve?

   - **Strategy:** Make lifelong growth a core part of your daily routine. Set aside time each day or week for activities that promote learning and self-improvement, whether it's reading, taking a course, or engaging in reflective practices like journaling or meditation. Seek out new challenges that push you out of your comfort zone and allow you to develop new skills or perspectives. By committing to lifelong growth, you ensure that you are constantly evolving, becoming the best version of yourself, and living a life of purpose and fulfillment.

Your journey, like Dawn's, is filled with moments that challenge and shape you. As you navigate your own path, remember that within every challenge lies the opportunity to deepen your purpose, build resilience, and embrace vulnerability as a source of strength. By cultivating gratitude and committing to lifelong growth, you can transform obstacles into stepping-stones toward a life of greater meaning and fulfillment. Just as Dawn did, you have the power to rise above your circumstances, continuously evolve, and leave a lasting impact on the world around you. Your story is still being written, and each day offers a new chance to live with purpose and grace.

To hear Dr. Dawn's story in her own words and to learn more about the causes she supports:

https://drivingchangepodcast.com/podcast/dr-dawn-mussallem-running-through-life-with-purpose-and-grace/

https://www.mayoclinic.org/biographies/mussallem-dawn-m-d-o/bio-20055494

https://donatelife.net/

CHAPTER 4

Hilary Phelps' Story

Finding the Right Swim Lane: A Journey of Recovery, Rediscovery and Hope

Hilary Phelps' journey is one of both extraordinary achievement and profound struggle, a story that navigates the complex terrain of success and self-worth. To the world, she was known as the older sister of the most decorated Olympian in history, Michael Phelps, and a world-class swimmer in her own right. But beneath the surface of her accomplishments, Hilary's life was marked by a battle that few saw—a battle that nearly consumed her before she found the strength to reclaim her life.

Hilary was the eldest of three siblings, and from a young age, she felt the weight of being the oldest in a family that revolved around excellence. Her parents, devoted and determined, had built their dream home in the woods, a sanctuary where Hilary and her siblings would roam, exploring nature with an unbridled sense of freedom. They spent their early years riding bikes through the forest, discovering the world around them, and cultivating a love for the outdoors. It was an idyllic childhood, one that seemed to lay the foundation for the greatness that would follow.

At the age of seven, the same year her brother Michael was born, Hilary's life took a pivotal turn. Her parents enrolled her in the local swimming club, and what began as a way to expend her boundless energy quickly revealed itself as something much more significant. Hilary had a natural talent for swimming, and it wasn't long before she was collecting trophies. But it wasn't just the thrill of competition that drove her—it was an insatiable hunger to be the best. When she won a third-place trophy, she looked at her mother and said, "This isn't enough. I want the big one." And with that declaration, Hilary embarked on a journey of relentless pursuit.

Swimming became the center of her life. By the time she was 11, Hilary was the fastest swimmer in the country in her event, a distance swimmer specializing in the mile. She wasn't just competing; she was dominating. Her parents, recognizing her potential, did everything they could to support her dreams. They sold their dream home and moved the family closer to Baltimore, where Hilary could train with the best. Her father would wake up at 3:45 every morning to drive her to practice, and her mother, a teacher, juggled her schedule to ensure Hilary could attend the best schools.

But as her swimming career soared, Hilary's inner world began to crumble. It started innocuously enough, with a simple comment from a group of girls at her new school. They said, "You have nice shoes." To anyone else, it might have seemed like a compliment, but to Hilary, it was a dagger. In that moment, those girls' voices became louder than her own. Despite her straight A's, despite being the fastest swimmer in the country, Hilary began to believe that she wasn't enough—not pretty enough, not smart enough, not good enough. It was as if all the confidence she had built in the pool evaporated in an instant.

This was the turning point, the moment when Hilary's life began to spiral. At the age of 14, she started drinking, seeking solace in alcohol to quiet the storm inside her. What started as a way to numb her insecurities quickly became an escape from the relentless pressure, she placed on herself. Hilary was a high achiever, driven by an internal force to be the best, but the expectations—both real and perceived—became too much to bear. The alcohol provided a temporary reprieve, a way to drown out the voices that told her she wasn't enough.

As the oldest child, Hilary also felt the pressure of being the role model, the one who had to have it all together. But behind closed doors, her life was anything but together. The drinking, which began as an occasional escape, soon took on a life of its own. By the time she entered college, Hilary was no longer just dabbling in alcohol—she was consumed by it. She drank to blackout, her nights lost in a fog of intoxication, and she convinced herself that this was normal. After all, wasn't college just one big party?

But while Hilary was excelling in the pool—earning a full scholarship to the University of Richmond and swimming competitively—her life outside of it was unraveling. She surrounded herself with people who drank like she did, who wouldn't question her habits because they shared them. She became adept at hiding her addiction, creating elaborate routines to disguise her drinking from others. She would rotate liquor stores to avoid suspicion, make excuses for her behavior, and build relationships with people who wouldn't hold her accountable. Her life revolved around alcohol—drinking, recovering from drinking, and thinking about the next drink. It was a cycle she couldn't break, even as it began to take a toll on her physically and emotionally.

Hilary's family life was also fraught with challenges. Her parents' divorce, which occurred around the same time she began drinking, left her feeling even more isolated and lost. But with her siblings rising in the swimming world—her sister Whitney becoming the third-fastest swimmer in the world at 14 and her brother Michael on the cusp of greatness—Hilary felt overshadowed, struggling to find her place. The drinking escalated as Hilary tried to cope with the growing sense of inadequacy. She was a blackout drinker, waking up each morning with no memory of the night before, her body racked with pain from the alcohol she had consumed. But no matter how much she drank, it was never enough. The alcohol that once provided an escape was now trapping her in a life she didn't recognize. Yet, the thought of quitting was terrifying because alcohol had become so intertwined with her identity. Without it, who was she?

By the time Hilary reached her mid-twenties, the effects of her drinking became impossible to ignore. She was in pain—both physically and emotionally—and she began to realize that alcohol was no longer the friend it once was. It wasn't helping her escape; it was trapping her in a life she didn't want. The relationships she had, the jobs she held, everything was affected by her addiction, yet she wasn't ready to confront it. The thought of quitting was terrifying because alcohol had become so intertwined with her identity. Without it, who was she?

It wasn't until she was 29 that Hilary finally reached her breaking point. After years of denial, she could no longer ignore the damage that alcohol was doing to her life. She sought help, entering a treatment program and beginning the long, difficult journey to sobriety. It was in treatment that she learned to manage her emotions, to confront the internal chaos that had driven her to drink. She surrounded herself with a new community, people who were also trying to stay sober one day at a time. She learned to replace her destructive habits with healthier ones—drinking tea instead of wine, practicing mindfulness, and focusing on self-care.

Hilary's recovery was not a linear process. There were setbacks, moments when she wanted to give up, but she kept moving forward. She embraced the tools she was given in treatment and built a new life for herself, one that wasn't defined by alcohol or the expectations of others. Over time, she regained her reading comprehension, her love of knowledge, and most importantly, her sense of self-worth.

Today, Hilary is a beacon of hope for others who are struggling with addiction. She has turned her pain into purpose, using her story to help other women find their own paths to healing. She created "The Right Room," a space where women can come together, share their stories, and support one another in their journeys of recovery and self-discovery. She understands that healing is not just about quitting alcohol; it's about rebuilding your life from the ground up, finding your voice, and learning to love yourself again.

Hilary's story is not just about overcoming addiction; it's about finding the courage to be who you truly are, even when the world is telling you to be someone else. In sharing her story, she invites us all to reflect on our own lives, to consider the ways in which we might be hiding from our true selves, and to find the strength to step into the light. Her message is one of hope, of the belief that no matter how far you've fallen, there is always a way back, and that the journey to healing begins with the simple act of believing in yourself.

Reflection: Embracing Transformation and Growth

As we reflect on Hilary Phelps' incredible journey, it's clear that her life is a powerful testament to the resilience of the human spirit and the profound transformation that can occur when we choose to face our inner struggles head-on. From being a world-class athlete to nearly losing herself to addiction and then finding her way back to a life of purpose and meaning, Hilary's story is not just about overcoming addiction—it's about living a life that is true to who we are, no matter the obstacles we face. Her journey invites us to consider how we, too, can navigate our own lives with the same courage, honesty, and commitment to growth.

Now, I invite you to take a moment to reflect on your own journey. The story of Hilary's life offers us more than just inspiration; it provides us with a roadmap—one that can help us navigate the complexities of our own experiences. Let's explore five key life lessons and strategies that we can glean from Hilary's journey, each offering a powerful tool to help us embrace transformation and growth.

1. **Listen to Your Inner Voice: The Guide to Authenticity**

   One of the most profound moments in Hilary's story is when she allowed the voices of others to drown out her own, leading her to believe she wasn't enough despite her accomplishments. The moment we stop listening to our inner voice—the voice that knows our true worth—is the moment we start to lose ourselves.

   - **Reflection:** Are you in tune with your inner voice, or are external opinions shaping your self-worth? What steps can you take to ensure that your inner voice remains the loudest and most guiding force in your life?

   - **Strategy:** Make a habit of tuning into your inner voice. This can be through journaling, meditation, or simply spending time alone in reflection. Ask yourself: "What do I truly believe about myself?

Am I letting external opinions shape my self-worth?" Regularly check in with yourself to ensure that your inner voice remains the loudest and most guiding force in your life.

2. **Surround Yourself with the Right People: The Influence of Community**

Hilary's healing began when she sought out a community that supported her sobriety and her true self. The people we surround ourselves with have a tremendous impact on our lives, either lifting us up or dragging us down.

- **Reflection:** Who are the people in your life that influence your journey? Are they lifting you up or holding you back? How can you seek out a community that supports your growth and well-being?

- **Strategy:** Evaluate the relationships in your life. Are the people around you lifting you up, or are they dragging you down? Seek out those who encourage your growth, who hold you accountable, and who inspire you to be the best version of yourself. If necessary, don't be afraid to distance yourself from those who don't support your journey toward wholeness.

3. **Embrace Your True Identity: The Path to Inner Peace**

Hilary's struggle with alcohol was, in many ways, a struggle with her identity. It was only when she embraced who she really was that she began to find peace. Embracing our true identity is essential for living a life of authenticity and fulfillment.

- **Reflection:** Do you truly know who you are at your core? Are you living in alignment with your true identity, or are you trying to fit into a mold that doesn't suit you?

- **Strategy:** Take time to explore and embrace your true identity. Who are you at your core? What do you value? What are your strengths and passions? Write these down and revisit them often. The more you understand and accept who you are, the less likely you are to feel the need to escape into behaviors or habits that aren't aligned with your true self.

4. **Recognize the Power of Habits: The Building Blocks of Change**

Hilary's addiction to alcohol was a deeply ingrained habit, but she also learned that positive habits could be just as powerful in her recovery. The habits we cultivate, whether positive or negative, have a profound impact on our lives.

- **Reflection:** What habits are shaping your life? Are they helping you move forward, or are they holding you back? How can you replace negative habits with positive ones that support your growth?

- **Strategy:** Identify the habits in your life that are either helping or hindering your progress. Start small by changing one habit at a time. If there's a behavior you want to eliminate, replace it with something positive. For example, if you find yourself turning to unhealthy coping mechanisms when stressed, replace that habit with something nurturing, like going for a walk, practicing deep breathing, or talking to a supportive friend.

5. **Never Underestimate the Power of Hope and Purpose: The Fuel for Transformation**

Even in her darkest moments, Hilary never completely lost hope. She found strength in the belief that her life could be different and held onto that hope until she was ready to act on it. Today, she uses her story to inspire and help others, finding purpose in her pain.

- **Reflection:** How do you cultivate hope in your life? Do you focus on the possibilities rather than the limitations? How can you find purpose in your struggles and use them as a catalyst on your journey?

- **Strategy:** Cultivate hope in your own life by focusing on the possibilities rather than the limitations. When faced with challenges, ask yourself: "What can I learn from this? How can I grow? And most importantly, how can I use my experiences to help others?" By finding purpose in your struggles, you transform them from burdens into lessons on your journey.

Hilary Phelps' story is a reminder that we all have the capacity to change, to heal, and to live a life that is true to who we are. It's not an easy journey—it requires honesty, courage, and the willingness to face uncomfortable truths about ourselves. But as Hilary's story shows, the rewards of this journey are profound. By listening to your inner voice, surrounding yourself with the right people, embracing your true identity, cultivating positive habits, and holding onto hope, you can create a life that not only fulfills you but also inspires others.

As you move forward, take these lessons to heart, and apply them in your own life. Remember that your story is still being written, and you have the power to shape it in ways that reflect the best of who you are. Your journey, like Hilary's, can be one of transformation and growth, and it begins with the choices you make today.

To hear Hilary tell her story and learn more about her platform today:

https://drivingchangepodcast.com/podcast/empowerment-and-purpose-a-world-class-swimmers-journey-through-addictionhilary-phelps/

https://www.hilaryphelps.com/

CHAPTER 5

The Unexpected Walk-On

Ben Malcolmson's Journey of Purpose

Ben Malcolmson never planned to play football at USC. In fact, the idea seemed as ridiculous to him as it might to anyone else. Growing up, Ben was just like any other kid, full of dreams and unsure of which path to take. He loved sports, but by the time he reached college, he had resigned himself to the fact that his role in sports would be from the sidelines—not as a player, but as a storyteller.

Ben's journey at USC began as a student journalist. After floundering through his freshman year, changing majors four times, he finally found his niche in journalism. Writing came naturally to him, and what better way to stay close to the sports he loved than to cover them? Soon enough, Ben found himself as the beat writer for the USC football team, one of the most storied and successful programs in college football history. This was the mid-2000s, the era of USC dominance under Coach Pete Carroll. The Trojans were at the height of their glory, winning 34 straight games and appearing in three consecutive national championships.

For three years, Ben reported on this dynasty, living vicariously through the athletes he wrote about, never imagining he would one day share the field with them. As his senior year approached, Ben faced the challenge every journalist knows well: how to write that final, unforgettable story. He wanted to go out with a bang, to leave his mark on the Daily Trojan with a story that would be remembered.

It was then that Ben noticed a small announcement about upcoming walk-on tryouts for the football team. Walk-ons, those unsung heroes who rarely see the field, intrigued him. They were the underdogs, the guys who showed up day after day without scholarships, without glory, just for the love of the game. Ben saw potential in their stories, and he decided to write about them. But then, a wild idea sparked in his mind: What if he didn't just write about the tryout? What if he went through it himself?

The thought was absurd. Ben hadn't played football since the fifth grade, and now he was considering trying out for the number one team in the country. He laughed at the thought, imagining the comedic potential of a student journalist getting his butt kicked by future NFL players. The story would practically write itself.

So, Ben signed up. He showed up to the tryout with no expectations, ready to turn his humiliating experience into gold for his final article. He ran a 40-yard dash for the first time in his life, struggled through drills he had never even heard of, and tried not to make a complete fool of himself. The whole experience felt like a bizarre dream—a dream that, by the next morning, would be nothing more than a funny story to tell.

But the next morning, something unimaginable happened. Ben received a phone call that would change his life. The voice on the other end of the line informed him that he had made the team. At first, Ben thought it was a joke. How could it not be? He had gone to the tryout as a lark, never expecting to be taken seriously. But the voice insisted, "You made the team. Come check in."

Ben's mind raced. This had to be one of Coach Pete Carroll's legendary pranks. Coach Carroll was known for his playful spirit, his ability to keep the atmosphere light even in the high-stakes world of college football. Surely, this was one of his elaborate jokes. But Ben had to know for sure, so he jumped on his bike and pedaled furiously to the football building.

As he climbed the steps, Ben's heart pounded. He braced himself for the punchline, for the hidden cameras that would capture his moment of humiliation. But when he reached Coach Carroll, there was no prank. Instead, Coach Carroll looked at Ben with a wide grin and said, "You can catch the ball. You can run fast. We want you on the team. Are you in?"

Ben stood there, stunned. Was this really happening? One of the best college football coaches in the country was offering him a spot on the team. The absurdity of it all washed over him, but so did a wave of excitement. He had no idea what the future held, but he knew he couldn't say no to this once-in-a-lifetime opportunity.

From that moment, Ben's life took a turn he never could have predicted. He was no longer just a student journalist; he was a wide receiver on the number one football team in the country. But the transition was anything but smooth. On his first day of practice, Ben was a fish out of water. He didn't know how to put on his pads, didn't understand the playbook, and felt like he was speaking a foreign language every time a coach barked out instructions. But Ben was determined. Despite the odds, despite the steep learning curve, he kept showing up, kept pushing himself, kept believing that somehow, this was where he was meant to be.

However, just as Ben started to find his footing, he was hit with another challenge. During spring practice, he dislocated his shoulder. The injury was severe enough to require reconstructive surgery, with a recovery time of nine months—longer than the entire football season. Ben was devastated. After finally starting to believe that he could contribute to the team, his dreams were seemingly crushed. But Ben wasn't ready to give up. He threw himself into his rehab, clinging to the hope that he could return to the field before the season ended.

Miraculously, Ben's recovery defied all expectations. Within four and a half months—half the expected time—his shoulder was healed, and he was cleared to play. It felt like a sign, a confirmation that he was indeed meant to be on that team for a reason.

As Ben rejoined his teammates, he felt a renewed sense of purpose. But this purpose wasn't about football. From the moment he made the team, Ben had felt a deep, unshakable sense that he was there for something more. He wasn't just a wide receiver; he was on a mission, a mission that he couldn't yet fully understand.

Ben was a man of faith, and he believed that God had placed him on that team for a reason. He tried to share his faith with his teammates, inviting them to Bible studies and prayer groups, but no one came. He tried to be a source of encouragement, a friend who would listen, but he didn't see any fruit from his efforts. As the season progressed, Ben began to doubt his purpose. Maybe he was just supposed to be a football player after all. Maybe he had been wrong about having a higher calling.

But just when Ben was on the verge of giving up, he felt a nudge on his heart, a new idea. With the Rose Bowl approaching, he decided to place a Bible in every teammate's locker. It was a bold move, one that he hoped would finally reveal the reason he was on the team. On Christmas Eve, Ben snuck into the locker room, carefully placing a Bible in each locker. As he did, that old sense of purpose flared up again. This was it. This was why he was there.

But when the team returned from the holiday break, Ben's heart sank. Instead of finding his teammates reading the Bibles, he found the locker room floor littered with torn pages. The Bibles had been ripped apart, discarded like trash. Ben was crushed. Once again, he felt like a failure, convinced that he had missed his purpose.

Then, as the season came to a close, tragedy struck. Mario, the team's kicker and one of Ben's closest friends, passed away unexpectedly. The grief was overwhelming, compounded by Ben's lingering sense of failure. But at Mario's funeral, Ben saw something that shook him to his core: atop Mario's casket was one of the Bibles he had placed in the locker room. Ben had no idea what it meant, but in that moment, he felt a small glimmer of hope, a reminder that maybe his efforts hadn't been in vain after all.

After graduating, Ben stayed on at USC, working with Coach Carroll. When Coach Carroll took the head coaching job with the Seattle Seahawks in 2010, he asked Ben to join him in Seattle. It was a new chapter in Ben's life, one where he could start fresh, free from the weight of his perceived failures at USC.

But as fate would have it, Ben's past wasn't done with him. Shortly after moving to Seattle, Ben attended an informational meeting for Young Life, a high school mentorship program. As he walked into the room, he was stunned to see a familiar face—Taylor, the punter from the USC football team. The two reconnected, and as they talked, Taylor shared a story that left Ben speechless.

Taylor told Ben how, during their final week on the team, he had seen one of the Bibles in the locker room and tossed it aside, not thinking much of it. But then, as he was leaving for practice, he heard a voice in the back of the locker room. It was Mario, sitting in his locker, thumbing through the Bible, asking, "What is this?"

Taylor, who hadn't opened a Bible in years, felt a sudden urge to turn back. He sat down next to Mario and spent the next 45 minutes explaining the words of Jesus to him. For the rest of that week, Mario couldn't put the Bible down. He had encountered God in a profound way, starting a relationship with God that Taylor had the privilege of witnessing. Just a few days later, Mario passed away.

As Taylor shared this story, Ben felt tears welling up. He had thought his efforts with the Bibles had been a total failure, but the reality was, his efforts had changed lives in ways he could never have imagined. Ben realized that even when we can't see the results, our actions can have a profound impact on others.

This realization led Ben to write his book, **"Walk On,"** a powerful testament to the idea that we all have a purpose, even when we can't see it. The title is a double entendre—referring to his experience as a walk-on wide receiver and as an encouragement to keep walking on in faith, trusting that there is a purpose for each of us, even in the most unexpected places.

Ben's story is a reminder that we may never fully understand the impact of our actions, but that doesn't mean our efforts are in vain. His journey from student journalist to football player, from despair to hope, and from failure to fulfillment, shows us that purpose is often found in the most surprising places. It's a story of faith, perseverance, and the power of unseen purpose—a story that encourages us all to keep walking on, no matter where life takes us.

Reflection: Pursuing the Unseen Purpose

Ben Malcolmson's unexpected journey from student journalist to walk-on wide receiver at USC is a story of faith, perseverance, and the power of unseen purpose. His experience reminds us that life's detours often lead us to places we never imagined, but it's in these unexpected paths that we often discover our true purpose. Ben's story is a powerful testament to the idea that our actions, even when they seem insignificant, can have a profound impact on others.

1. **Embrace the Unexpected Detours: Discovering New Paths**

   Ben never intended to play football at USC; his goal was to write about the experience of trying out for the team. But when he unexpectedly made the team, Ben found himself on a journey that would reveal a deeper purpose in his life. His story teaches us that life rarely goes according to plan, and our greatest opportunities often arise from the most unexpected detours.

   - **Reflection:** Ben's journey encourages us to remain open to life's surprises. Often, the paths we didn't plan for lead us to the most fulfilling destinations. Reflect on the unexpected turns your life has taken. How have these detours shaped who you are today? What opportunities might you discover if you embraced these changes rather than resisted them?

   - **Strategy:** The next time life throws you a curveball, resist the urge to dismiss it as a mistake or inconvenience. Instead, lean into it. Ask yourself, "What new possibilities does this present? How can I grow from this experience?" Document these reflections in a journal, noting how each detour has contributed to your personal growth. By embracing life's unexpected turns, you open yourself to new possibilities that may lead to your true purpose.

2. **Listen to the Nudge on Your Heart: Trusting Inner Guidance**

Throughout his time on the football team, Ben felt a persistent nudge on his heart—a sense that he was there for a reason beyond just playing football. Even when his efforts seemed to bear no fruit, Ben continued to follow this inner guidance, which eventually led to a profound impact on his teammates. His story reminds us that the quiet, persistent feelings we have often point us toward our deeper purpose.

- **Reflection:** Ben's experience teaches us the importance of trusting the inner nudges that guide us. These feelings, though often subtle, can lead us to our true calling if we have the courage to follow them. Reflect on the times when you've felt a strong inner pull toward something, even if it didn't make logical sense. What might happen if you trusted these feelings and acted on them?

- **Strategy:** Pay attention to the recurring thoughts, feelings, or ideas that tug at your heart. These nudges are often clues to your deeper purpose. Instead of brushing them aside, take them seriously. Start by taking small steps in the direction of these feelings, whether it's pursuing a new interest, reaching out to someone, or making a life change. Trust that these nudges are guiding you toward something meaningful and allow them to inform your decisions.

3. **Persevere Through Discouragement: Finding Strength in Faith**

Ben's journey wasn't without its challenges. From being sidelined by a shoulder injury to seeing his efforts to share his faith seemingly fail, Ben faced numerous setbacks that could have led him to give up.

However, he chose to persevere, driven by a deep sense of purpose and faith. His story is a powerful reminder that persistence in the face of discouragement often leads to the most profound rewards.

- **Reflection:** Ben's perseverance through discouragement is a testament to the power of faith and determination. Reflect on the times in your life when you've faced setbacks or felt like giving up. How did you respond? What might have happened if you had persisted? Ben's story encourages us to dig deeper into our faith and purpose when we face obstacles, trusting that our efforts are not in vain.

- **Strategy:** When you encounter setbacks, remind yourself that discouragement is often a sign that you're on the right path. Instead of giving in to doubt, dig deeper into your "why." Why did you start this journey? What keeps you going? Use these reflections as fuel to keep moving forward, even when the results aren't visible. Write down your reasons for persisting and revisit them when you feel discouraged. By maintaining your commitment, you open the door to unexpected breakthroughs.

4. **Trust That Your Actions Have Impact, Even When You Can't See It**

When Ben placed Bibles in his teammates' lockers, he initially thought his efforts had failed. It wasn't until later that he learned his actions had a profound impact, leading to life-changing experiences for some of his teammates. Ben's story highlights the importance of trusting that our actions, even when we can't see immediate results, can make a difference in ways we may never fully understand.

- **Reflection:** It's easy to feel disheartened when our efforts don't yield immediate results, but Ben's story reminds us that the impact of our actions often unfolds in ways we can't see. Reflect on a time when you felt your efforts were in vain, only to later discover their impact. How might your perspective change if you trusted that every positive action you take has the potential to create change, even if the results aren't visible?

- **Strategy:** Adopt a long-term perspective on the impact of your actions. Understand that the seeds you plant today may not bear fruit until much later. Trust that every positive action, every word of encouragement, and every step taken in faith has the potential to change lives in ways you may never see. Keep sowing good seeds, even when the soil seems barren. To reinforce this mindset, keep a journal where you document your positive actions and periodically reflect on how they might be making a difference, even if the outcomes aren't immediately apparent.

5. **Find Purpose in Serving Others: The True Measure of Success**

Ben's journey ultimately led him to discover that his true purpose was not about personal achievement, but about serving others. His efforts, which at first seemed fruitless, were actually laying the foundation for profound changes in the lives of those around him. Ben's story is a powerful reminder that our greatest fulfillment often comes from making a difference in the lives of others.

- **Reflection:** Ben's story challenges us to rethink our definition of success. It's not always about what we achieve for ourselves, but about the impact we have on others. Reflect on how you're using your skills, talents, and experiences to serve others. Are you focused more on personal success, or are you seeking ways to make a lasting impact on those around you?

- **Strategy:** Consider how your actions can contribute to the well-being of others. Identify opportunities where you can use your skills and resources to serve, whether it's through your work, volunteering, or simply offering support to someone in need.

Make service a core part of your life's purpose. To keep this at the forefront of your mind, set aside time each week to engage in activities that benefit others, and reflect on how these actions contribute to your overall sense of purpose and fulfillment.

To learn more about Ben and hear his story in his own words:

https://drivingchangepodcast.com/podcast/podcast-ben-malcolmson/

Ben's Book, Walk On:

https://www.amazon.com/Walk-Sidelines-My-Unlikely-Football-Following/dp/0735291276

CHAPTER 6

James Taggart's Story

The Journey From Imposter to Finding Real Identity

In the sweltering summer of 1974, in the city of Cincinnati, Ohio, a nine-year-old boy named James Taggart was thrust into a nightmare that would forever alter the course of his life. Picture a young boy, the youngest of eight siblings, waking up early on a Saturday morning with the simple task of delivering newspapers. It was a routine part of his life, part of the family paper route that had been passed down from sibling to sibling. But on that fateful morning, as James walked out to the garage to get his bike, he opened the door to a scene that would haunt him for decades.

What he saw in that dimly lit garage was not something a child should ever have to witness. Panic overtook him. He didn't fully comprehend what he was seeing, but his instincts knew. He ran as fast as his legs could carry him, screaming for his mom. In that moment of sheer terror, his mind couldn't process the truth—it was his dad, hanging lifelessly from the rafters, a victim of his own despair. James's father had taken his own life, plunging the entire family into a vortex of grief, confusion, and survival.

Imagine being nine years old and discovering that the person who was supposed to protect and guide you had chosen to leave this world in such a tragic way. For James, that moment marked the beginning of a long and arduous journey through a life filled with emotional landmines. The 1970s were a time when therapy and mental health were stigmatized, especially in cases of suicide. Society viewed such tragedies with suspicion and fear, often leaving families like James's isolated. There was no open conversation, no outreach for help—only a grim determination to survive.

James describes the next 25 years of his life as being in "survival mode," a state of constant vigilance and emotional repression. He didn't have the luxury of processing his grief; there was no time to heal. Instead, he buried his pain deep inside, where it festered and manifested in physical symptoms, particularly in his lower back. The weight of his unprocessed emotions, the anger, sadness, and fear, all settled in his body, causing chronic pain that would plague him for over two decades.

It's hard to fully grasp what it means to live in survival mode unless you've been there. It's a place where you're always on edge, where every decision is driven by the need to protect yourself from further harm. For James, this meant becoming fiercely determined and independent. He decided he would live life on his own terms, no matter how unbalanced that might make him. He threw himself into work, staying busy to avoid the quiet moments when his thoughts might catch up with him. He pursued financial success with a singular focus, believing that money would provide the security and control he so desperately craved.

But life, as it often does, has a way of showing us the futility of our plans. James's relentless drive to control everything around him only led to more pain and more problems. His back issues worsened, and by his mid-twenties, he was barely able to function. At 27, he was spending 10 to 15 minutes every morning just trying to stand up straight, his body racked with pain. Imagine the frustration, the humiliation, of being a young man in the prime of life, unable to do something as simple as lifting a five-pound object. For James, every day was a battle against his own body, a constant reminder of the unresolved trauma that lay beneath the surface.

By 1992, James had seen over 60 different doctors—chiropractors, physical therapists, back specialists, and more. He had tried every treatment available, from electrical stimulation to acupuncture, but nothing provided lasting relief. He was told he would need surgery, but he was too terrified to go through with it. Instead, he spent five weeks out of work, crawling on his hands and knees just to get to the bathroom. His life was deteriorating, his world shrinking as his physical limitations grew.

Then, in a twist of fate, James met a man named Tom Curtin, a martial arts instructor who had opened a healing center in Orlando. Tom had suffered a severe back injury himself and had healed through the movements he now taught. Desperate and with nothing left to lose, James decided to give it a try. Skeptical but hopeful, he attended one of Tom's demonstrations and saw something that gave him a glimmer of hope. Tom moved with ease, clearly free of the pain that had once debilitated him. Maybe, just maybe, there was something here for James, too.

It's important to pause here and consider what this moment must have felt like for James. After years of disappointment and despair, he was being offered a lifeline by a stranger who seemed to understand his pain. How many of us, when faced with our darkest moments, have yearned for someone to come along and show us a way out? For James, Tom was that person—a teacher, a guide, someone who had walked a similar path and emerged stronger on the other side.

James began working with Tom, learning movements that were completely different from anything he had tried before. In just three weeks, he experienced a breakthrough—he could get out of bed and stand up straight without pain. It was a small victory, but for James, it was monumental. The relief he felt was like a light at the end of a very long, dark tunnel. For the first time in years, he had hope.

But as anyone who has ever tried to overcome deep-seated issues knows, the journey was far from over. James was still driven by his old habits—pushing himself too hard, too fast, in a desperate bid to reclaim his life. Tom had his work cut out for him, constantly reminding James to slow down, to be patient. It was a lesson that took time to sink in, but slowly, James began to heal, not just physically but emotionally and mentally as well.

Over the next 22 months, James's life transformed. He corrected his back problems and found a new sense of balance. He became a martial arts instructor himself, teaching others the very techniques that had saved him.

The last time he experienced any significant back pain was in 1994. His journey from a young boy traumatized by his father's suicide to a man who had reclaimed his life and found his purpose was nothing short of extraordinary.

But James's story doesn't end there. While his physical pain had subsided, he still had to confront the emotional and psychological wounds that had driven him for so long. In many ways, he had been living as an imposter—projecting an image of confidence and success to the world, while inside he struggled with feelings of inadequacy and fear. It's a familiar story to many of us, isn't it? How often do we wear masks to hide our true selves, fearing that if people saw who we really were, they wouldn't accept us?

For James, the journey to authenticity was long and winding. After many years of focused effort on healing and releasing the past, he eventually found a unique way to embrace his true self—through impersonating someone else. It's ironic, isn't it? The man who had spent so many years feeling like an imposter would find success as a professional impersonator, embodying none other than James Bond, the epitome of cool, confidence, and charisma.

It started in 2006, when the first Daniel Craig Bond film, *Casino Royale*, hit theaters. People began to stop James on the street, remarking on his resemblance to the new 007. At first, he brushed it off. He didn't see the likeness himself—after all, how often do we really look at our own faces? But the comments kept coming, and by 2011, after three people approached him within 90 minutes to comment on his resemblance to Daniel Craig, James decided to explore the possibility of becoming a Bond impersonator.

James had no experience in modeling or acting, and the idea of putting himself out there in such a public way was a bit foreign. But he took the plunge, got some professional photos taken, and reached out to an agency that specialized in lookalike work. His first gig was a learning experience, but it opened his eyes to a whole new world. He met other impersonators, heard their stories, and realized that this could be more than just a hobby—it could be a career.

And so, James Taggart, who had spent so many years feeling like an imposter in his own life, became the world's premier James Bond impersonator. He traveled the world, working in seven countries and more than 30 states, meeting fascinating people and having incredible experiences along the way. The very thing that had once been a source of insecurity for him—his resemblance to someone else—became his greatest asset.

But James didn't just stop at being an impersonator. He had always dreamed of living on a sailboat, a dream that had been with him since childhood when he read stories of explorers like Magellan and Vasco da Gama. After years of designing a life to support this dream, James finally took the plunge. He started by taking sailing lessons, then spent eleven months on a sailboat through a timeshare-like program. Eventually, he bought his own catamaran, which he named H_2 007, and began using it for business purposes.

James has now fulfilled the goal of making that sailboat his permanent home. He plans to live a life of freedom and adventure, sailing from place to place, with no fixed address. It's a life that many of us can only dream of, but for James, it's the culmination of a lifelong journey—a journey from a young boy traumatized by his father's death to a man who has found his own path, his own peace, and his own purpose.

As we reflect on James's story, it's impossible not to be struck by the sheer resilience and determination that carried him through his darkest moments. His journey teaches us that no matter how deep our wounds, no matter how overwhelming our pain, there is always a way forward. It may not be easy, and it may take time—years, even decades—but healing is possible. We all have the capacity to transform our pain into purpose, to turn our struggles into strengths.

James's life is a testament to the power of personal empowerment. He has lived life on his own terms, refusing to let his past define him or his pain limit him. He has embraced his true self, imperfections and all, and in doing so, he has found a way to not only survive but to thrive.

Reflection: Living Authentically and Purposefully

As we reflect on James's compelling journey, it's clear that his life is a powerful reminder of the importance of acknowledging our pain, embracing our true selves, and living with purpose. James's story isn't just about overcoming personal struggles—it's about the transformation that occurs when we stop merely surviving and start living authentically. His journey invites us to consider how we, too, can navigate our own lives with the same self-awareness, resilience, and intention.

The story of James's life offers us more than just inspiration; it provides us with a roadmap—one that can help us navigate the complexities of our own experiences. Let's explore five key life lessons and strategies that we can glean from James's journey, each offering a powerful tool to help us live more authentically and purposefully.

1. **Acknowledge Your Pain, Don't Ignore It: The Path to Healing**

   James spent years running from the pain of his past, burying it under layers of work, success, and control. His chronic back pain was a manifestation of the unresolved trauma he carried with him. James's journey teaches us that pain, whether physical, emotional, or psychological, must be acknowledged before healing can begin.

   - **Reflection:** What pain or unresolved emotions are you carrying with you? Are there areas in your life where you've tried to push through or ignore the signals your body or mind are sending you?

   - **Strategy:** Take a moment to acknowledge that pain and consider what it might be trying to tell you. Healing begins with recognition and acceptance. Allow yourself to explore your pain, perhaps through journaling, therapy, or simply sitting with your emotions, and consider what steps you can take to address it.

2. **Survival Mode Isn't Sustainable: Shifting to a Balanced Life**

   In the wake of his father's suicide, James entered survival mode—a state that kept him functioning but also kept him from truly living. Survival mode is not sustainable; it drains our energy and prevents us from experiencing true peace and fulfillment.

   - **Reflection:** Are you living in survival mode? Are you constantly on edge, trying to control every aspect of your life? How might this be affecting your well-being?

   - **Strategy:** Consider how you can shift from survival to a more balanced, sustainable way of living. Start by identifying one area where you can let go of control and trust in the process. Focus on small changes that can help you move from a state of mere survival to one of intentional living.

3. **Find Your Teacher: The Importance of Guidance**

   James's life began to change when he met his teacher, Tom Curtin, who introduced him to new ways of thinking and living. Often, we need mentors, guides, and teachers to help us navigate life's challenges and discover new paths forward.

   - **Reflection:** Who are the teachers in your life? Are there mentors, friends, or even authors who have guided you on your journey? If not, where might you find such guidance?

   - **Strategy:** Be open to the wisdom that others can offer and seek out those who have walked a similar path. Whether it's through formal mentorship, books, or conversations with trusted individuals, be willing to learn from those who can provide insight and support on your journey.

4. **Embrace Your Authentic Self: The Courage to Be Real**

For much of his life, James was an impersonator—not just in his role as a James Bond lookalike, but in the way he presented himself to the world. His journey to healing involved peeling back those layers and embracing his true, authentic self.

- **Reflection:** Are you living authentically, or are you wearing a mask to hide your true self? What would it mean for you to step into your authenticity, to embrace who you truly are without fear or shame?

- **Strategy:** Reflect on areas in your life where you might be holding back your true self. Consider how you can start to show up more authentically. This might involve having honest conversations, pursuing passions that resonate with your true self, or simply allowing yourself to be vulnerable in front of others.

5. **Live with Purpose, Not Just Survival: Creating a Life of Meaning**

As James transitioned from survival mode to a more purposeful way of living, he discovered new passions and joys. He learned that life isn't just about getting through the day; it's about living with intention and creating a life that reflects your true values.

- **Reflection:** Are you living with purpose, or are you simply surviving each day? What are the passions and dreams that you've put on hold? How can you start living more purposefully?

- **Strategy:** Take time to reflect on what truly matters to you, and consider how you can start living more purposefully. What small steps can you take today to align your life more closely with your values and passions? Whether it's pursuing a dream, cultivating a hobby, or simply making more intentional choices, start moving towards a life that reflects your true purpose.

James's story is a testament to the power of resilience, self-awareness, and the pursuit of authenticity. As you reflect on these lessons, remember that your journey is unique, and the path to healing and fulfillment is not always linear. Embrace the process, seek out support, and trust that you have the strength to overcome whatever challenges come your way.

As you move forward, take these lessons to heart, and apply them in your own life. Remember that your story is still being written, and you have the power to shape it in ways that reflect the best of who you are. Your journey, like James's, can be one of transformation and growth, and it begins with the choices you make today.

To hear James' story in his own words and to learn more about him:

> https://drivingchangepodcast.com/podcast/james-taggart-the-journey-to-authenticity/
>
> https://www.007now.com/
>
> https://www.007sailing.com

CHAPTER 7

Katie Spotz

Uncharted Waters: Pushing Limits and Defying Expectations

Katie grew up in a world where she was constantly chasing something. With two older brothers and a house full of their friends, she learned early on that if she wanted to keep up, she'd have to push herself harder. But it wasn't just about keeping up with the boys—Katie was driven by an insatiable curiosity. She was the kind of child who couldn't sit still, always climbing, always moving, always exploring. There was a spark within her, a need to see what was beyond the next hill, to understand the limits of what she could do.

But despite this boundless energy, Katie didn't see herself as an athlete. She tried her hand at the usual sports—soccer, basketball, baseball—but none of them clicked. She was present, physically, but mentally she was elsewhere. In baseball, you might find her picking flowers in the outfield, her mind far from the game. In soccer, she often lost track of which side of the field she was supposed to be on. Team sports just didn't resonate with her, and by the time she reached high school, she had firmly decided that athletics were not her path. In her mind, she simply wasn't cut out for it.

Katie's disinterest in sports reached its peak during her senior year of high school. She needed one last gym class to graduate, but she did everything she could to avoid it. She petitioned for waivers, tried to write papers in lieu of physical activity, and even considered getting a doctor's note. But there was no escape—she had to take the class. She chose a walking and running course, determined to put in as little effort as possible.

But life has a way of surprising us when we least expect it. One day, as she walked around the track, boredom set in. A thought crept into her mind, almost as if it had been waiting there all along: *Could I run a mile?* It seemed like a simple question, but for Katie, it was profound. She had never done it before and wasn't sure if she could. But the curiosity that had always driven her as a child stirred once more, and she decided to find out.

The run wasn't easy. She struggled to pace herself, her legs felt heavy, and her lungs burned. But step by step, she pushed forward. And then, before she knew it, she had done it. She had run an entire mile. In that moment, something shifted inside her. The story she had told herself for so long—that she wasn't an athlete, that she couldn't do it—began to unravel. If she had been wrong about this, what else had she been wrong about? What other limits had she accepted without question?

That mile was a revelation. It wasn't just a physical accomplishment; it was a challenge to her entire self-concept. Katie began to wonder: *If I can do this, what else can I do?* The seed of curiosity that had been dormant in her throughout high school was now sprouting into something new—a desire to explore her own potential, to push beyond the limits she had set for herself.

With each new mile, Katie's confidence grew. She began to test herself, seeing how far she could go. One mile turned into two, then three, and each one felt like a victory, a step further into uncharted territory. By the time she hit double digits, the idea of running a marathon—a feat that once seemed impossible—began to take shape in her mind.

But Katie wasn't content with just running a marathon. She had discovered something in herself, a drive to see just how far she could go, and she wanted to explore it fully. After completing her first marathon, she set her sights on an even greater challenge: an Ironman triathlon. But in true Katie fashion, she didn't approach it the way most people would.

Instead of following a traditional training plan, Katie decided to test her cycling ability in the most extreme way possible—by signing up for a 3,300-mile bike ride across the United States. The ride, which would raise funds for the American Lung Association in memory of her grandmother, was a massive undertaking. And in typical Katie style, she didn't even own a bike when she made the decision. But that didn't deter her. She bought a bike, trained hard, and set off on a journey that would take her from one coast of the United States to the other.

The ride was grueling. There were days when the road seemed endless, when the fatigue in her legs threatened to overwhelm her. But Katie had discovered something powerful during her run—that the mind is as important as the body in endurance sports. She pushed through the pain, through the exhaustion, knowing that each mile brought her closer to her goal. And when she finally completed the ride, she knew that she had unlocked something within herself—a resilience, a determination, and a capacity for endurance that she hadn't known she possessed.

With the bike ride behind her, Katie's curiosity led her to her next challenge: swimming. But not just any swim—Katie decided to tackle the 325-mile Allegheny River. This was no ordinary swim; it was an endurance test on a grand scale, requiring months of preparation and training. But for Katie, the physical challenge was only part of the story. As she trained, she became increasingly aware of a global issue that resonated deeply with her: the water crisis.

Katie's awareness of the water crisis began during her time living in Australia, where a severe drought was affecting everything from daily life to the national economy. As she learned more, she was struck by the stark reality that over a billion people worldwide lacked access to clean water. Growing up on the shores of the Great Lakes, Katie had always taken water for granted. The realization that so many people lived without this basic necessity hit her hard.

Her decision to swim the Allegheny River became more than just a personal challenge—it became a mission. She dedicated the swim to raising awareness and funds for clean water projects, transforming her physical journey into a vehicle for change. The swim was tough, pushing her body to its limits, but knowing that she was swimming for a cause larger than herself gave her the strength to keep going.

Katie's biggest challenge came in 2010 when she decided to row solo across the Atlantic Ocean. It was an idea that seemed almost impossible, even to her. Rowing 3,000 miles across open water, alone, with no support boat and no resupply, was a test of endurance, courage, and mental fortitude on a scale she had never experienced before.

Her boat, a 19-foot vessel designed specifically for ocean rowing, became her world for 70 days. It was a small, cramped space, with a coffin-sized cabin that served as her living quarters, and a tiny cockpit for storage. The boat was designed to self-right if it capsized, but that was small comfort when facing the vast, unpredictable Atlantic Ocean.

As she set off from the coast of West Africa, Katie knew she was embarking on a journey that would test every part of her being. The first days were filled with excitement and adrenaline, but as the reality of the challenge set in, so did the isolation. She was completely alone, thousands of miles from the nearest human being. The ocean was both her companion and her adversary, offering moments of serene beauty and terrifying power.

The physical challenges were immense. Katie rowed for hours each day, her muscles aching, her hands blistered. Sleep was a luxury she rarely afforded herself—two hours at a time, if she was lucky, in a cabin that often felt like a sauna. There were days when the exhaustion seemed too much to bear, when the idea of continuing felt overwhelming. But Katie knew that she couldn't think about the entire journey; it was too vast, too daunting. Instead, she focused on rowing one mile at a time. Just one mile. And then another. And another.

There were moments of deep despair, especially when she reached the halfway point. Instead of feeling relief, she was overcome with the realization that she was only halfway through. The thought of doing it all over again, of rowing another 1,500 miles, seemed impossible. She broke down in tears, overwhelmed by the enormity of the task before her. But even in that moment of despair, Katie found a way to keep going. She made a conscious decision to let go of the big picture and focus only on the next stroke, the next mile.

It was in these moments of extreme vulnerability that Katie's true strength emerged. The journey across the Atlantic was as much a mental and spiritual challenge as it was a physical one. She learned to embrace the solitude, to find peace in the rhythm of the oars and the endless horizon. She discovered a resilience within herself that she hadn't known existed—a resilience that came not from overpowering the ocean, but from surrendering to it, from finding a way to coexist with the vastness around her.

Throughout her journey across the Atlantic, Katie's thoughts often returned to the cause that had become so dear to her: the global water crisis. The stark contrast between the endless water beneath her and the scarcity faced by so many people around the world weighed heavily on her mind. She had come to realize that her endurance challenges could be more than just personal tests; they could be powerful tools for raising awareness and making a tangible difference.

Katie's work for clean water didn't end with her Atlantic row. She continued to push the limits of her endurance, running ultra-marathons, completing more Ironman triathlons, and even rollerblading across the Florida Keys—all to raise funds and awareness for clean water projects. Each challenge was driven by a deep sense of purpose, a commitment to using her abilities to help others.

But Katie's journey wasn't without setbacks. During one of her most recent challenges—an ultra-marathon on rollerblades—she took a severe tumble, tearing her ACL and MCL. For someone who had pushed her body to such incredible limits, the injury was a harsh reminder of her own vulnerability. But even as she faced surgery and the long road to recovery, Katie's indomitable spirit remained unbroken. She saw the injury not as an end, but as a new beginning, an opportunity to come back stronger than ever. And come back she did as she set and completed a goal of going from zero to 26.2miles (marathon) within 6 months after knee surgery! She's now currently working towards completing the first 100 mile run post-surgery and has been able to already complete a 70-mile run so she's well on her way!

Based on how fast Katie sets and accomplishes these amazing goals, it's quite likely that any updates on future goals I report here will already be outdated by the time you're reading this, however, Katie's next BIG goal is "A Long Way 4 Water" a 3,100-mile cycle across the continent of South America from the Atlantic to Pacific Coast, as a continuation of the Row for Water. As of writing this, she has completed 1,500-miles and in a matter of weeks is set to complete the final 1,600-miles from Brazil to Lima, Peru. You can follow her progress here: https://katiespotz.com/alongway4water

is her most ambitious yet: a 200-mile ultra-marathon, just months after recovering from her knee surgery. It's a challenge that would daunt even the most seasoned athletes, but for Katie, it's another step in her journey—a journey defined not by the limits of her body, but by the boundless potential of her spirit.

Katie Spotz's story is one of relentless curiosity, courage, and commitment. From her early days as a curious child to her incredible feats of endurance, Katie has consistently defied the limits of what she believed was possible. Her journey is not just about physical accomplishments; it's about the power of the human spirit to push beyond fear, to embrace the unknown, and to make a difference in the world.

Through her work, Katie has shown that we are all capable of more than we realize. Her story is a reminder that the limits we perceive are often self-imposed, and that with courage, determination, and a sense of purpose, we can achieve the extraordinary. Katie's journey is far from over, and as she continues to push the boundaries of what's possible, she inspires us all to do the same.

Reflection: Five Key Life Lessons from Katie Spotz's Journey

Katie Spotz's story is more than an inspiring tale of endurance and adventure; it's a profound lesson in what it means to push the limits of human potential. As we reflect on her journey, we can extract powerful life lessons that apply to all of us, no matter where we are on our own paths. These lessons are not just about physical endurance; they are about the mindset, courage, and purpose that drive us forward. Let's explore five key takeaways from Katie's story that can help guide you on your journey.

1. **Embrace Curiosity: The Gateway to Growth**

   Katie's entire journey began with a simple question: *Could I run a mile?* This curiosity, this willingness to ask questions and explore the unknown, was the spark that ignited a lifetime of incredible achievements. Curiosity isn't just about seeking knowledge; it's about challenging the status quo and asking, *What if?*

   - **Reflection:** What areas of your life have you accepted as fixed or unchangeable? What would happen if you allowed yourself to be curious, to explore new possibilities? The first step toward growth is often as simple as asking a question. Don't be afraid to wonder, to imagine, and to explore. It might just lead you to the next great adventure in your life.

   - **Strategy:** Start small. Identify one area of your life where you've been complacent or where you've accepted limitations. Ask yourself, *What if I tried something different?* Take a small step toward exploring that curiosity. It could be as simple as picking up a new book, trying a new hobby, or asking a question you've been afraid to ask. Let your curiosity guide you toward growth.

2. **Face Your Fears: The Key to Unlocking Potential**

   Katie's journey was marked by moments of intense fear and doubt. From running her first mile to rowing across the Atlantic, she constantly faced challenges that seemed insurmountable. But instead of shying away, she leaned into those fears, knowing that on the other side of fear is often the greatest growth.

- **Reflection:** What fears are holding you back? Are there challenges in your life that you've avoided because they seem too daunting? Often, the things we fear the most are the very things that can unlock our potential. Facing those fears head-on is the key to breaking through barriers and discovering what we're truly capable of.

- **Strategy:** Identify a fear that has been holding you back. It doesn't have to be a monumental fear; it could be something as simple as the fear of failure, rejection, or change. Take a step toward confronting that fear. Break it down into smaller, manageable actions, and tackle it one step at a time. Remember, just as Katie did, that you don't have to conquer the entire mountain at once—focus on taking that first step.

3. **Celebrate Small Wins: The Foundation of Momentum**

One of the most powerful aspects of Katie's journey was her ability to celebrate each small victory along the way. Whether it was running her first mile or completing a marathon, she didn't wait until she reached the finish line to acknowledge her progress. Each small win fueled her motivation to keep going.

- **Reflection:** How often do you celebrate your progress? Do you acknowledge the small wins, or do you only focus on the end goal? Celebrating small victories is crucial for building momentum and staying motivated. It's a reminder that every step forward, no matter how small, is a step in the right direction.

- **Strategy:** Set small, achievable goals on your journey, and celebrate each one as you accomplish it. These could be daily, weekly, or monthly milestones. The key is to recognize your progress and allow yourself to feel proud of what you've achieved. This practice builds confidence and creates the momentum needed to tackle even bigger challenges.

4. **Find Your Purpose: The Fuel for Endurance**

Katie's journey wasn't just about personal achievement; it was deeply connected to a greater purpose. Her mission to provide clean water to those in need gave her a sense of meaning that fueled her endurance through even the toughest challenges. Purpose is what transforms an ordinary goal into something extraordinary.

- **Reflection:** What is your greater purpose? What drives you beyond personal success or accomplishment? Finding a purpose that resonates deeply with you can provide the motivation and resilience needed to overcome obstacles and push through difficult times.

- **Strategy:** Reflect on what matters most to you. What causes, values, or goals resonate with you on a deeper level? How can you align your actions and goals with this greater purpose? When you connect your journey to a purpose larger than yourself, you tap into a wellspring of motivation and strength that can carry you through any challenge.

5. **Break It Down: The Power of Incremental Progress**

When Katie felt overwhelmed by the enormity of rowing 3,000 miles across the Atlantic, she broke it down into smaller, manageable tasks—rowing one mile at a time. This approach allowed her to stay focused and keep moving forward, even when the overall goal seemed impossible.

- **Reflection:** Are you feeling overwhelmed by a big goal or challenge? It's easy to get lost in the magnitude of what we want to achieve. But by breaking it down into smaller, more manageable tasks, we can make steady progress without feeling overwhelmed.

- **Strategy:** Take a big goal or challenge you're facing and break it down into smaller, actionable steps. Focus on completing one step at a time, rather than getting caught up in the enormity of the entire task. This approach not only makes the goal more achievable but also helps you build confidence as you see your progress unfold.

Katie's journey is a powerful reminder that we are all capable of more than we realize. By embracing curiosity, facing our fears, celebrating small wins, finding our purpose, and breaking down big goals into smaller steps, we can unlock our full potential and achieve extraordinary things. Each of these lessons offers a strategy for navigating life's challenges with courage and resilience. As you reflect on your own journey, consider how you can apply these lessons to move closer to your goals and live a life of purpose and fulfillment.

To hear Katie's story and learn more about her journey:

https://drivingchangepodcast.com/podcast/katie-spotz-discovering-your-own-potential/

https://katiespotz.com/

https://donorbox.org/alongway4water

CHAPTER 8

Jordan Jonas

The Adventure of a Lifetime

The vast, open plains of Idaho stretch out under an endless sky, a place where nature's rhythms dictate the pace of life. Here, a young boy named Jordan Jonas was unknowingly being prepared for a future that would test the very limits of human endurance. Born in Illinois but raised on the rugged farmlands of Idaho, Jordan's early life was a crucible where he learned the hard-won lessons of resilience, patience, and the deep, silent wisdom of the earth.

On the farm, Jordan grew up surrounded by the simple but profound realities of life. The cycles of planting and harvest, the quiet dignity of hard work, and the stark, unyielding beauty of the natural world shaped his young mind. He wasn't just learning how to tend the land; he was learning how to survive. Every day brought a new lesson in self-reliance, whether it was mending a broken fence, tending to the animals, or simply standing in awe of a sunrise that painted the sky in hues of pink and gold. These early years were the foundation of a life that would later unfold in extraordinary ways.

But Jordan's story was never destined to remain within the boundaries of that Idaho farm. There was a restlessness in him, a yearning for something more, something that couldn't be found in the predictable rhythms of farm life. This hunger for the unknown would take him on a journey across continents, through cultures, and into the most extreme conditions a human being can endure. It would test not just his physical strength, but his mental fortitude, his ability to adapt, and his capacity to find meaning during hardship.

The roots of Jordan's resilience run deep, tracing back to a family history marked by unimaginable suffering and enduring strength. His grandparents, Assyrians who survived the Armenian genocide, carried with them stories of loss and survival that would later become a touchstone for Jordan's own life. His grandmother, a woman of extraordinary fortitude, witnessed the deaths of six of her seven siblings during a brutal death march through the desert. She and her sister, barely more than children themselves, were forced to drag their mother to her feet, urging her to keep moving when all seemed lost. They were eventually saved by the mercy of a British army camp, a glimmer of hope in a landscape of despair.

Jordan's grandfather's story was no less harrowing. As a young boy, he watched in horror as his father was burned alive, his sister disappeared, and he was left an orphan in a world that seemed to have lost its humanity. Taken in by Jesuit priests in Iraq, he found a new kind of family among those who had also lost everything. These stories of survival and resilience were the bedrock upon which Jordan's own life would be built. They weren't just tales from a distant past; they were the lifeblood of a lineage that had survived against all odds.

Although Jordan never met his grandparents—they passed away before he was born—their legacy lived on in the stories told by his father and his 11 aunts and uncles. These weren't stories of bitterness or hatred, but of love, endurance, and an unshakable resilience. The ability to rise above suffering, to refuse to be defined by tragedy, was a lesson that Jordan absorbed deeply. It was a perspective that would guide him through the many trials he would face in his own life.

By the time Jordan was 18, the quiet life of the farm could no longer contain him. There was a fire burning within him, a need to explore, to push the boundaries of his own experience. This need led him and his brother to a life that few would dare to choose—a life riding the freight trains that crisscrossed the country, living the hobo lifestyle that had all but disappeared in the modern world.

For Jordan, this wasn't just about the thrill of adventure; it was about freedom. The freedom to wake up each day without a plan, to let the wind carry him to places he had never seen, to live by his wits and on his own terms. There was something deeply intoxicating about the open road, about the endless possibilities that lay just beyond the horizon. But this freedom came with its own set of challenges. There were nights spent sleeping under the stars, with nothing but the sound of the train's wheels clattering against the tracks. There were days when food was scarce, when the uncertainty of where the next meal would come from was a constant companion.

One morning, as they rumbled along in the belly of a freight train, Jordan's brother awoke with a dead owl on his head—a bizarre and unsettling omen that set the tone for the day. Later, they were spotted by railroad workers, who pointed and shouted, their gestures unmistakable: they were about to be caught. As the train slowed, Jordan and his brother tried to make a break for it, but they were quickly apprehended and hauled off to jail. It was a stark reminder that the freedom they sought was always balanced on the edge of danger.

But these experiences only deepened Jordan's resolve. He wasn't just running from something; he was running towards something—a deeper understanding of himself and the world around him. The open road was his classroom, and every challenge, every hardship was a lesson in resilience, in adaptability, in the simple, unadorned truth of what it means to be human.

Jordan's insatiable curiosity and his willingness to embrace the unknown would soon lead him far from the familiar landscapes of Idaho and the American railways. His journey took him to the remote wilderness of Siberia, where he would live among the native reindeer herders—a people whose way of life had remained largely unchanged for centuries.

The path that led Jordan to Siberia was as winding and unpredictable as the rivers that cut through the frozen tundra. It began with a desire to do something meaningful, to contribute to a cause greater than himself. He traveled to Russia to help build an orphanage, but he quickly realized that he didn't want to live in the relative comfort of the mission. He wanted to immerse himself in the culture, to live as the Russians lived, to experience the world as they did.

Through a series of chance encounters, Jordan met a fur trapper who lived among the native peoples of Siberia. This man, impressed by Jordan's spirit and determination, invited him to come north, to learn the ways of the trappers and the reindeer herders. Jordan accepted without hesitation, knowing that this was the kind of experience he had been seeking all his life.

At the age of 25, Jordan found himself deep in the Siberian wilderness, surrounded by a landscape that was both beautiful and unforgiving. The fur trapper dropped him off at a remote camp, introducing him to his cousins—nomadic reindeer herders who lived in teepees and moved with the seasons. And then, without ceremony, the trapper left, leaving Jordan to fend for himself among people who spoke a language he barely understood and lived a life he could hardly imagine.

For three years, Jordan lived with these reindeer herders, learning their ways, adapting to their harsh environment, and discovering a way of life that was both alien and deeply familiar. The Siberian wilderness was a world apart from anything Jordan had known, yet it felt like home. The reindeer herders taught him the ancient skills of survival—how to trap, how to navigate the land, how to live in harmony with nature. It was a life stripped down to its essentials, where every day was a test of resilience and resourcefulness.

Jordan thrived in this environment, not just because of his physical skills, but because of his mindset—his ability to adapt, to embrace the unknown, and to find joy in the simplicity of life. There were moments of danger, like the time he nearly drowned while crossing a raging river on the back of a reindeer. But there were also moments of profound connection, moments when Jordan felt a deep kinship with the land and the people who called it home.

When Jordan was cast on the History Channel's show *Alone*, it was as if all the experiences of his life had been leading up to this moment. The show was a test of survival, pure and simple; contestants were dropped into the wilderness with nothing but a few tools and their wits, and the last person standing would win. For Jordan, this wasn't just a competition; it was a chance to prove to himself what he was truly capable of.

For 77 days, Jordan lived alone in the wild, enduring the harsh elements, the physical challenges, and the mental strain of complete isolation. But unlike many of the other contestants, Jordan wasn't daunted by the prospect of being alone. To him, this was just another chapter in a life that had been defined by resilience and adaptability.

From the moment he was dropped into the wilderness, Jordan's survival instincts kicked in. He scouted the area, assessing the resources available to him. He built a shelter, fashioned tools from the materials around him, and set traps to catch food. But the physical challenges were only part of the equation. The real test was mental—the ability to stay focused, to stay positive, and to keep moving forward even when the odds seemed insurmountable.

Jordan knew that the key to survival was staying busy. He set goals for himself each day, whether it was improving his shelter, foraging for food, or creating new tools. He also knew the importance of humor—of finding joy in the little things, even when everything seemed challenging. He entertained himself by staging elaborate skits, reenacting famous battles, and doing impressions of characters from his favorite movies. These moments of levity were a way of keeping his spirits up, of reminding himself that, no matter how dire the situation, there was always something to laugh about.

But there were also moments of deep introspection. Alone in the wilderness, with nothing but his thoughts for company, Jordan was forced to confront the darker corners of his mind. He thought about his family, about the choices he had made in his life, about the legacy he wanted to leave behind. These moments of solitude were both challenging and enlightening, giving Jordan a deeper understanding of himself and his place in the world.

As the days turned into weeks, and the weeks into months, Jordan's resilience was tested again and again. There were moments when he doubted himself, when the isolation and the hunger threatened to overwhelm him. But each time, he found the strength to push through, drawing on the lessons of his past, on the resilience of his ancestors, and on the simple, unshakable belief that he could survive. He did, in fact, survive; outlasting all the other contestants and being declared the official winner of season six of the show!

Jordan's story is not just a tale of survival. It's a testament to the power of the human spirit, to the resilience that lies within each of us, waiting to be called upon in times of need. It's a story of a man who was never content to live a life of comfort and ease, who sought out challenges not to prove something to the world, but to discover something within himself. It's a story of a life lived with purpose, with intention, and with a deep, abiding love for the natural world.

In the end, Jordan's journey is a reminder that life's greatest adventures often begin at the edge of our comfort zone. It's a call to embrace the unknown, to trust in our own resilience, and to find joy in the journey, no matter how rugged the path may be.

As we close this chapter of Jordan's story, we are left with a deep sense of admiration for a man who has lived a life that few of us could imagine, but who reminds us that the greatest adventures are often the ones we choose to embark on ourselves. Jordan Jonas is not just a survivor; he is a living example of what it means to live life fully, to embrace the challenges that come our way, and to find meaning in even the most difficult of circumstances.

Reflection: Lessons from the Wilderness of Life

As we step back from Jordan Jonas's incredible journey, it's impossible not to feel a sense of awe at the resilience and adaptability he demonstrated in the face of seemingly insurmountable challenges. But more than that, his story serves as a mirror, reflecting at us the potential we each hold to face our own wilderness—whether that wilderness is physical, emotional, or metaphorical. Let's take a moment to reflect on five key life lessons from Jordan's story and how we can apply these lessons to our own lives.

1. **Embrace Your Roots: The Power of Resilience**

Jordan's resilience was not just a trait he developed in the wilderness; it was deeply rooted in his family's history. The stories of his grandparents, who survived the Armenian genocide and went on to live lives marked by love and strength, were a constant reminder of the power of resilience. Their ability to rise above unimaginable suffering and continue forward with grace and courage was passed down to Jordan, shaping his own approach to life's challenges.

- **Reflection:** Take a moment to think about the legacy of resilience in your own life. Who are the people in your family or community whose stories of survival and perseverance inspire you? How can you draw strength from their experiences when you face your own trials? Consider writing down these stories or sharing them with others as a way to honor that legacy and keep it alive within you.

- **Strategy:** When faced with a difficult situation, pause and remind yourself of the resilience that runs through your veins. Reflect on the challenges your ancestors or role models overcame and use their strength as a foundation for your own. This perspective can shift your mindset from one of defeat to one of empowerment.

2. **Seek Adventure: The Value of Stepping Outside Your Comfort Zone**

Jordan's decision to ride the rails, live among reindeer herders in Siberia, and eventually compete on *Alone* was driven by a desire to push beyond the boundaries of comfort and familiarity. Each new adventure was a step into the unknown, a chance to discover what he was truly capable of. This willingness to embrace the unknown is what allowed Jordan to grow, to learn, and to thrive in environments that would have broken many others.

- **Reflection:** How often do you step outside your comfort zone? Are there areas of your life where you've allowed fear or complacency to keep you from pursuing new experiences? What would happen if you took a leap into the unknown, trusting in your ability to adapt and grow?

- **Strategy:** Challenge yourself to take one action this week that pushes you out of your comfort zone. It doesn't have to be as extreme as riding the rails or living in Siberia—it could be as simple as trying a new activity, having a difficult conversation, or tackling a project that intimidates you. Each time you step into the unknown, you build the muscle of courage and expand your capacity for growth.

3. **Develop Mental Fortitude: Winning the Battle Within**

One of the most striking aspects of Jordan's story is his ability to maintain mental strength in the face of isolation and adversity. The wilderness stripped away all distractions, leaving him alone with his thoughts. But instead of succumbing to fear or despair, Jordan used this time to build his mental resilience, to stay focused on his goals, and to keep moving forward.

- **Reflection:** In our modern world, it's easy to distract ourselves from the mental battles we face. We fill our days with noise and activity, avoiding the quiet moments where our deeper thoughts and fears might surface. But true mental fortitude is developed in those quiet moments, when we confront our inner demons and emerge stronger.

- **Strategy:** Set aside time each day for quiet reflection or meditation. Use this time to check in with yourself—your thoughts, your emotions, your fears. Instead of avoiding the difficult or uncomfortable feelings, face them head-on. Acknowledge them, process them, and then let them go. Over time, this practice will build your mental resilience and prepare you for whatever challenges come your way.

4. **Adapt and Overcome: The Importance of Flexibility**

When Jordan arrived in the wilderness, many of his initial plans were quickly thwarted by the harsh reality of his environment. But instead of giving up, he adapted. He let go of his original expectations and created new strategies that allowed him to thrive in the conditions he was given. This ability to adapt—to bend without breaking—is a crucial skill in both survival and life.

- **Reflection:** Think about a time when things didn't go according to plan. How did you react? Were you able to adapt, or did you struggle to let go of your original expectations? How might your life be different if you approached challenges with a mindset of flexibility and adaptability?

- **Strategy:** The next time you're faced with an unexpected obstacle, take a step back and assess the situation. What are your new realities? What resources do you have at your disposal? Instead of clinging to your original plan, ask yourself, "What can I do with what I have?" Embrace the change and use it as an opportunity to grow and evolve.

5. **Find Joy in the Journey: The Role of Humor and Perspective**

Even in the most challenging circumstances, Jordan found ways to keep his spirits high. He entertained himself with skits, laughed at the absurdities of his situation, and maintained a sense of humor that kept him grounded. This ability to find joy in the journey, no matter how difficult, is a powerful tool for resilience.

- **Reflection:** How often do you allow yourself to laugh, even during stress or difficulty? Do you take time to find the humor in life, to appreciate the small moments of joy that can make all the difference? Humor and perspective are not just coping mechanisms; they are vital components of a resilient mindset.

- **Strategy:** Make it a point to inject humor and joy into your daily life. Surround yourself with people who make you laugh, watch a funny movie, or simply take a moment to smile at the little things that bring you happiness. Remember that no matter how serious life gets, there's always room for laughter.

Jordan's story is a powerful reminder that we all have the capacity to survive and thrive, no matter what challenges life throws our way. But survival is not just about enduring hardship; it's about embracing the journey, learning from each experience, and emerging stronger on the other side.

As you reflect on these lessons, think about your own life. What wilderness are you facing? What challenges are testing your resilience, your adaptability, your mental strength? And most importantly, how can you use these lessons to navigate your journey with courage, grace, and a sense of adventure?

Remember, the wilderness is not a place to fear—it's a place to discover who you truly are. Embrace it, and you may find that the person you become is stronger, wiser, and more resilient than you ever imagined.

To hear Jordan tell his story and learn more about his journey:

https://drivingchangepodcast.com/podcast/jordanjonas/

https://www.jordanjonas.com/

CHAPTER 9

Johnny Sirpilla's Story

From Pain to Purpose: Finding Strength in Surrender

Sometimes, life has a way of testing us in ways we never imagined, pushing us to the edge of what we think we can endure. Johnny Sirpilla's story is one of those that, when you hear it, it makes you pause, take a breath, and really consider the depth of human resilience. It's a story that pulls at the heartstrings, but also one that pushes us to reflect on how we handle our own darkest moments, how we find strength when everything seems lost.

Johnny grew up with the kind of foundation many of us can only dream of—loving parents who set high expectations, who taught him and his sisters that their name was something to be honored, that integrity was non-negotiable, and that hard work was the path to any success worth having. Johnny was raised to understand that life wouldn't always be easy, but with the right mindset, anything could be overcome. This was the bedrock of resilience his parents unknowingly instilled in him—resilience that would later be tested in ways none of them could have anticipated.

In those early years, life was good—really good. Johnny was a star in his career, and his wife, Susan, was thriving in hers. They were the quintessential power couple, succeeding in every way that society tells us matters. But behind the scenes, a different story was unfolding—one of private struggle and deepening despair. For years, Johnny and Susan faced the silent, isolating battle of infertility. They wanted a family more than anything, and like so many of us who set our sights on a goal, they believed that if they just worked hard enough, they could make it happen.

But this wasn't something they could control. Can you imagine the frustration? The helplessness? Here were two people who had achieved so much, yet they were powerless against this invisible foe. Infertility isn't something we talk about enough—it's one of those silent struggles that can feel so isolating, so deeply personal. And for Johnny and Susan, it was compounded by the expectations they had set for themselves. They were planners, doers, achievers. But no amount of planning or doing could make this work the way they wanted.

And then, finally, a glimmer of hope. After countless procedures and emotional roller coasters, they found out they were pregnant. Not with one child, but with three—triplets. It must have felt like the universe was finally rewarding them for all their perseverance, for all their hope and faith. But life, as it so often does, had other plans. Just when they thought their dreams were about to be realized, they were instead handed their greatest nightmare. Nicholas, Mary, and Peter, their three precious babies, passed away shortly after birth.

It's hard to even imagine the depth of that pain, isn't it? The kind of loss that takes your breath away, that leaves you questioning everything you thought you knew about life, about fairness, about faith. For Johnny and Susan, this wasn't just a loss—it was a shattering of their world. And yet, what stands out most in Johnny's story isn't just the tragedy itself, but how he and Susan responded to it.

In moments like these, many of us would be tempted to turn inward, to let grief pull us apart. But Johnny and Susan made a different choice—a choice that speaks to the incredible strength of their relationship and their individual characters. The night before the funeral, an acquaintance who had also lost a child showed up at their door. They brought with them a simple but profound piece of advice: grieve together, not apart. This advice became a lifeline for Johnny and Susan, helping them to lean on each other rather than let the grief tear them apart.

Imagine that for a moment. You're in the darkest place of your life, and someone reaches out with just the right words, just the right wisdom. It's a reminder that even in our most isolated moments, we're not alone. And it's a reminder of the power of community, of reaching out to others even when we're not sure what to say or do. Sometimes, just showing up is enough.

But this wasn't just about getting through the funeral—it was about a commitment they made to each other, to not let this break them. With the help of their psychologist, Dr. Barbara Fordyce, they began the process of cognitive restructuring—a method of reframing their thoughts, of finding meaning in their loss rather than letting it define them. It wasn't about forgetting Nicholas, Mary, and Peter—it was about honoring their brief lives by choosing to live with gratitude and hope, even in the midst of unimaginable pain.

How many of us could do that? It's easy to talk about resilience in theory, but when you're in the thick of it, when every part of you just wants to give up, that's when the real test begins. Johnny and Susan didn't just survive—they chose to thrive, to rebuild their lives in a way that honored their children and their love for each other.

But the story doesn't end there. As if to prove that life is full of unexpected turns, Johnny and Susan were blessed with another miracle—adoption. They welcomed their son, Beau, into their lives, and not long after, despite all the medical advice and the odds, Susan became pregnant again. Twice. Bella and Stone, their two "miracle" children, were born prematurely but healthy, bringing their family full circle. But even this chapter in their lives was fraught with roadblocks as they suffered through several failed adoptions, a miscarriage with twins and the very real life threatening pregnancies for Susan that ultimately led to the births of Bella and Stone. It's a testament to the idea that even when life knocks you down, there's always the possibility of something beautiful just around the corner.

Johnny's experience also changed him as a leader. After the loss of his children, he couldn't return to a work environment that was negative, that allowed for pettiness and division. He realized that if he was going to survive—if he was going to truly live—he needed to create a culture at work that reflected the values he held dear. He couldn't allow the toxic behavior that so often creeps into businesses. Instead, he fostered an environment of support, empathy, and teamwork.

Think about that for a moment. How many of us carry our personal struggles into our professional lives? How often do we let our own pain or stress color the way we lead, the way we treat others? Johnny understood that his employees were also dealing with their own challenges, their own hidden battles. And as a leader, he felt a responsibility to create a workplace where people could not only succeed professionally but also find the support they needed to face whatever was happening in their personal lives.

In his book, *Life is Hard, But I'll Be Okay: The Power of Hope, Emerging Through Pain and Learning to Live with Gratitude,* Johnny shares the principles that guided him through these dark times. It's a book that isn't just about surviving loss—it's about turning that loss into something meaningful, about finding a deeper sense of purpose during the pain. Johnny's life is a powerful reminder that resilience isn't just about bouncing back—it's about moving forward with faith, with hope, and with a renewed commitment to live a life of gratitude, no matter what comes our way.

As we reflect on Johnny's story, it's impossible not to see the parallels in our own lives. We all face challenges—some big, some small—but what we do with those challenges is what defines us. Johnny chose to use his pain as a catalyst for growth, to find light in the darkness. And in **doing so, he offers us all a roadmap for how we might do the same.**

Reflection: Embracing Resilience and Redefining Strength Through the Lens of Loss

As we reflect on Johnny's profound journey, we see that his life embodies the power of resilience, perspective, and the ability to find meaning even in the face of unimaginable loss. Johnny's story isn't just about surviving hardship—it's about transforming pain into purpose and leading with empathy and compassion. His journey invites us to consider how we, too, can navigate our own lives with the same strength, grace, and intentionality.

Now, I invite you to take a moment to reflect on your own journey. The story of Johnny's life offers us more than just inspiration; it provides us with a roadmap—one that can help us navigate the complexities of our own experiences. Let's delve deeper into five key life lessons and strategies that we can glean from Johnny's journey, each offering a powerful tool to help us embrace resilience and redefine our strength.

1. **The Power of Perspective: Reframing Adversity**

   Johnny and Susan's journey through infertility and the loss of their triplets could have easily led them down a path of bitterness and despair. The weight of such a loss might have crushed anyone's spirit, leading to a life overshadowed by grief and regret. However, instead of being consumed by their sorrow, Johnny and Susan made the conscious choice to reframe their experience. They chose to see it not as an end, but as a beginning—a painful, heartbreaking beginning that offered the possibility of growth, healing, and transformation. This shift in perspective didn't diminish their pain, but it allowed them to find meaning within it, turning their grief into a foundation for future strength.

   - **Reflection:** How do you respond to adversity? When faced with life's challenges, do you allow them to cloud your perspective, or do you actively seek out the lessons they offer? Consider how shifting your focus from "Why me?" to "What now?" could transform your approach to hardship. What might happen if you viewed your challenges as opportunities for growth rather than insurmountable obstacles?

   - **Strategy:** The next time you encounter adversity, consciously work to shift your perspective. Ask yourself, "What can I learn from this? How can this pain propel me forward rather than hold me back?" By choosing to see challenges as catalysts for growth, you can find meaning and purpose even in the darkest of times. This doesn't mean ignoring your pain, but rather, it means embracing it as a vital part of your journey toward becoming stronger and more resilient.

2. **Grieving Together, Not Apart: The Strength in Connection**

One of the most poignant aspects of Johnny's story is how he and Susan chose to grieve together. Grieving is an intensely personal experience, and it can often drive people into isolation. The rawness of pain, the vulnerability of sorrow, can make us want to retreat, to hide from the world and even from those closest to us. However, Johnny and Susan made the brave choice to face their grief together. This decision required immense courage and vulnerability, but it also fortified their bond, allowing them to support each other in ways that solitary grieving could never achieve. Through shared tears, open conversations, and mutual support, they were able to find strength in each other's presence.

- **Reflection:** How do you handle grief and loss? Do you tend to isolate yourself, thinking that your pain is yours alone to bear? Or do you allow others to share in your grief, to walk alongside you in your darkest moments? Imagine how your experience of grief might change if you leaned into your relationships instead of retreating from them. What could be the impact of opening up, of letting others help carry your burden?

- **Strategy:** Consider how you can share your grief with those you trust. Allow them to support you through the healing process, recognizing that vulnerability can be a source of strength rather than weakness. Grieving together not only lightens the emotional load but also deepens the connections that give our lives meaning. This shared experience can create a powerful bond that strengthens relationships, fostering a deeper sense of connection and understanding.

3. **Resilience Is a Choice: The Power to Keep Moving Forward**

Johnny's journey is a powerful illustration that resilience is not something we are simply born with—it is a choice we make in the face of adversity. After the devastating loss of their triplets, Johnny and Susan could have easily succumbed to despair, allowing their grief to paralyze them. But instead, they made the conscious decision to rebuild their lives, step by step. Resilience isn't about being unbreakable; it's about refusing to stay broken. It's about choosing to rise each time you fall, to keep moving forward even when the path ahead is steep and treacherous.

- **Reflection:** How do you respond to challenges in your life? When you are faced with hardship, do you allow it to define you, or do you choose to define your response to it? Reflect on the moments when you've been knocked down—how did you get back up? What does resilience look like in your life, and how can you cultivate it further in the face of ongoing struggles?

- **Strategy:** Recognize that resilience is a daily choice. It's about taking small, actionable steps that allow you to keep moving forward, even when it feels difficult. When faced with hardship, remind yourself that you have the power to choose how you respond. Focus on what you can control, break down overwhelming tasks into manageable pieces, and celebrate small victories along the way. Each day, choose resilience—not because the journey is easy, but because the destination is worth the struggle.

4. **The Importance of Cognitive Restructuring: Shaping Your Thoughts**

One of the most practical tools Johnny and Susan used to navigate their grief was cognitive restructuring—a psychological technique that involves challenging and changing unhelpful thoughts. In the midst of their grief, it would have been easy for them to succumb to thoughts of hopelessness and despair. But instead, they actively worked to reshape their thinking, to challenge the narrative that their lives were defined by loss. Cognitive restructuring isn't about denying the reality of pain, but about finding a way to see beyond it, to recognize that even amid suffering, there is room for hope, growth, and new beginnings.

- **Reflection:** How do your thoughts shape your experience of challenges? Are your thoughts helping you move forward, or are they holding you back? Consider the power of your internal dialogue—are you telling yourself a story of defeat, or are you open to the possibility of growth and healing? What might happen if you challenged and reframed those thoughts?

- **Strategy:** Practice cognitive restructuring by identifying unhelpful thoughts and challenging them with a more balanced perspective. Instead of simply forcing yourself to think positively, aim for realistic thinking—acknowledge the pain, but also recognize the potential for growth. Ask yourself, "Is this thought based on fear, or is it grounded in reality? What evidence do I have to support or refute this thought?" By reframing your thoughts in this way, you can open yourself up to finding hope and meaning in your situation, even when it seems elusive.

5. **Leading with Empathy and Compassion: Transforming Leadership**

Johnny's experience of loss didn't just change him as a person—it transformed his approach to leadership. Before his loss, Johnny may have led with efficiency, with a focus on results and productivity. But after experiencing such profound grief, he realized the importance of creating a culture of empathy, compassion, and support in the workplace. He understood that leadership isn't just about guiding people to achieve goals; it's about caring for the whole person, recognizing their struggles, and creating an environment where they feel valued and supported.

- **Reflection:** How do you lead in your life, whether in your work, family, or community? Do you create environments where others feel supported, or do you prioritize other values, such as efficiency or control? Consider the impact of leading with empathy and compassion—how might it change the lives of those around you? How might it change you?

- **Strategy:** Consider how you can lead with more empathy and compassion in your daily life. Whether it's in the workplace, at home, or in your community, strive to create spaces where others feel valued and supported. This might mean taking the time to listen more, to understand the challenges others are facing, or to offer help and encouragement without judgment. Remember that leading with your heart doesn't just improve the lives of those you lead—it enriches your own life as well, fostering deeper connections and a greater sense of purpose.

Johnny Sirpilla's story is one of immense pain, but it's also one of incredible hope and transformation. It reminds us that no matter how hard life gets, we have the power to choose our response. We can choose to grieve together, to find strength in our relationships, to reframe our thoughts, and to lead with empathy. As you move forward in your own life, take these lessons to heart. Life will bring challenges—some small, some great—but how you respond to those challenges is what will ultimately define you. Johnny's story shows us that even in the darkest times, there is always hope, always the possibility of something beautiful just around the corner. The question is, will you be ready to embrace it?

To hear Johnny's story and learn more about his journey:

https://drivingchangepodcast.com/podcast/johnny-sirpilla-dealing-with-life-and-finding-perspective/

https://www.amazon.com/gp/product/B0B45DKBCZ

CHAPTER 10

Linda Cliatt-Wayman

So What, Now What? Transforming Challenges Into Change

In the heart of Philadelphia, in a neighborhood where hope seemed to be an elusive dream, a young girl named Linda Cliatt-Wayman was raised by a woman who refused to let the darkness of their surroundings define their lives. Linda's mother was a beacon of resilience, a woman who knew that the world was bigger than the impoverished streets of Judson Street, where they lived. She might not have seen it herself, but she had a vision—a vision that her daughters would rise above the limitations imposed upon them by society, poverty, and circumstance.

Imagine growing up in a place where every day was a battle for survival. Where the violence, the drugs, and the hopelessness that permeated the streets were the only reality you knew. But for Linda and her two sisters, their mother's unwavering belief in a better future was the thread of light that pierced through the thick fog of despair that hung over their community. This was a woman who, despite her own struggles and lack of resources, was determined to show her daughters that there was more to life than the narrow confines of Judson Street.

Every Sunday, after church, Linda's mother would gather her girls around the table. There were no family vacations, no trips to the amusement park or the beach. But what she did have were the bus schedules for the city. She would lay them out on the table and ask her daughters, "Which bus route are we going to ride today?" It was her way of saying, "Let's go see the world." It was a simple, yet profound act of defiance against the limitations that poverty placed on their lives. They would ride those buses from one end of the city to the other, exploring neighborhoods far removed from the hardships of their own. Along the way, their mother would point out houses and say, "Linda, if you want to live in a house like that, you have to go to school." She didn't have all the answers, but she knew that education was the key to unlocking the door to a better life.

What an incredible foresight for a woman who herself had never seen what lay beyond her own neighborhood. She didn't have the means to take her daughters on grand adventures, but she knew that showing them there was more to the world was crucial. She knew that for her daughters to aspire to something greater, they first had to see that "something greater" existed.

Linda's mother was more than just a parent; she was a change agent long before the term became popular. In an era when desegregation was just beginning, and many parents in the community were too afraid to send their children out of the neighborhood, Linda's mother made a bold choice. When the school district came knocking, offering to bus children to better schools in more affluent areas, every other parent said no. They were too fearful of the unknown, of sending their children to a place where they might not be welcomed. But not Linda's mother. She was the only parent who said yes, believing that exposure to diversity and different environments was essential for her daughters' growth.

The ridicule and criticism she faced from her community must have been immense. It's easy to imagine the whispers, the judgmental looks, the isolation she felt from her neighbors. But she stood firm in her belief that there was more out there for her girls. She knew that this one choice—this seemingly small decision—could alter the course of their lives. And she was right.

Linda's story could have ended there, with her simply becoming another statistic of inner-city Philadelphia. But her mother's decision to bus her out of the neighborhood exposed her to a different world—one where education was valued, where learning was a priority, and where the possibilities seemed endless. This experience planted the seeds of a lifelong passion in Linda, a passion for education and for helping children who, like her, grew up in circumstances that seemed designed to crush their spirits.

When Linda reached high school, her mother gave her the freedom to choose where she wanted to go. Linda chose the neighborhood high school, a decision that would forever change her. What she saw there shocked her—the stark inequities in education, the difference between what she had experienced in the schools she was bussed to and the reality of the schools in her own community. The contrast was jarring, and it ignited a fire within her. She knew that her life's work would be dedicated to closing that gap, to ensuring that all children, no matter where they came from, had access to a quality education.

Linda's journey took her from the classrooms of Philadelphia to the role of principal, and eventually to one of the most challenging positions anyone could imagine—leading a school labeled as one of the most dangerous in America. Strawberry Mansion High School was more than just a school; it was a battleground. For five consecutive years, it had been on the persistently dangerous list. The environment was chaotic, and many believed the children there were beyond saving. They were labeled as incorrigible, lost causes, products of their environment with no hope of redemption.

But Linda didn't see it that way. She never viewed any child as a lost cause. She understood, perhaps better than anyone, that the environment was the problem, not the kids. When she walked into Strawberry Mansion for the first time, she knew that the real challenge wasn't just to change the students—it was to change the very fabric of the school itself.

One of the most poignant moments in her early days at Strawberry Mansion came during a town hall meeting with the students. She stood before them, laying out an elaborate discipline system, complete with rewards for good behavior and consequences for breaking the rules. But then, a student stood up in the back of the room and said, "Ms. Wayman, why are you telling us about your rules and consequences? Don't you know we do what we want to do?" In that moment, Linda realized the depth of the problem. These children had been so neglected, so traumatized by their environment, that they truly believed they could come to school and do whatever they pleased without consequence.

It wasn't just that the school was dangerous; it was that the students had no concept of what a real school should be. To them, Strawberry Mansion was just a place to hang out, a place where the chaos of their lives continued without interruption. But Linda was determined to change that. She knew that before she could teach them anything, she had to create an environment where they felt safe, where they felt loved, and where they believed that their future could be different.

Changing that environment was no easy task. Linda faced death threats, resistance from the students, and fear from her staff. There were days when the danger was so palpable that she wondered if she would make it home. But she pressed on, driven by her belief that every child deserves a chance. She knew that if she could show these kids that she truly cared . . . that she was committed to them, they would begin to trust her. And that trust was the key to unlocking their potential.

It's easy to say you care, but proving it—day in and day out, especially in the face of such adversity—is another matter entirely. But Linda did it. She showed up every day, she held her staff accountable, and she implemented systems that slowly but surely began to change the school's culture. She listened to the students, and when they asked for something—like a football team that the school hadn't had in 50 years—she made it happen. Little by little, she earned their trust. And once she had that, everything began to change.

One of Linda's most powerful phrases—"So what, now what?"—became a mantra not just for her staff, but for the students as well. When faced with the overwhelming problems that seemed insurmountable, Linda would say, "So what? Now what are we going to do about it?" It was her way of pushing past the despair, of refusing to be paralyzed by the enormity of the challenges before her. It was a call to action, a reminder that no matter how bad things were, there was always something that could be done to make them better.

And it worked. By focusing on creating a safe learning environment, by implementing one system at a time, Linda and her team were able to do what seemed impossible—they got Strawberry Mansion off the persistently dangerous list. But more than that, they gave the students something they had never had before: hope. The transformation didn't happen overnight, and it wasn't easy. But it was real, and it was lasting.

For Linda, the ultimate goal was always about more than just improving test scores or getting off a list. It was about helping these children see that they were more than the labels society had placed on them. It was about showing them that, like her mother had shown her all those years ago, there was more to life than the narrow world they had known.

Linda's story is a testament to the power of one person to make a difference. Her mother's choice to bus her out of the neighborhood set her on a path that would lead her to change the lives of countless children. And now, Linda continues that legacy, not just through her work in the schools, but through her nonprofit, Currently Trending, which aims to increase graduation rates and break the cycle of poverty for young people in her community.

As Linda likes to remind her students every day, "If nobody told you they loved you today, you remember I do, and I always will." It's a simple phrase, but it carries with it the weight of a lifetime of love, commitment, and belief in the potential of every child.

So, to everyone listening, take a moment to think about the people in your life who have been that light for you. Who has shown you that there is more to life than the narrow world you've known? And then ask yourself, how can you be that light for someone else? Because, as Linda has shown us, it only takes one choice, one person, to change the course of a life forever.

Reflection: The Power of Purpose, Vision, and Unyielding Love

Linda Cliatt-Wayman's story is more than just an account of transforming a school—it's a powerful testament to the impact one person can have when driven by purpose, vision, and unyielding love. Her journey, from a young girl exploring the world beyond Judson Street with her mother to a principal who redefined education for students in some of Philadelphia's most challenging schools, serves as a profound reminder that true change begins with belief—belief in oneself, belief in others, and belief in a better future.

As we reflect on Linda's journey, there are key lessons and strategies that stand out—lessons that not only shaped her life but also offer us guidance on how to approach our own challenges, whatever they may be. Let's dive deeper into these lessons and explore how they can serve as a blueprint for our own lives.

1. **Purpose as a Driving Force: The Foundation of Impact**

    Linda's story begins with her mother's unwavering belief that there was more to life than what they saw on Judson Street. This belief wasn't just a fleeting thought; it was a deeply ingrained conviction that life held greater possibilities, and this conviction planted a seed of purpose in Linda that would guide her throughout her life. Purpose is not just a goal; it's the underlying reason that fuels our actions and decisions. It's the "why" behind everything we do, especially when the road is difficult, and the challenges seem insurmountable.

    - **Reflection:** What is your driving purpose? What is the deeper "why" that motivates you each day? Purpose is the cornerstone of resilience and perseverance. When challenges arise, it's your sense of purpose that will keep you moving forward. Consider the areas of your life where you feel most driven—what is the purpose that fuels these actions?

    - **Strategy:** Take time to clearly identify and articulate your purpose. Write it down, reflect on it, and let it guide your decisions. Purpose gives you the resilience to keep going when the road ahead is tough. It's the anchor that holds you steady amidst life's storms.

Whether your purpose is centered around your career, your family, or a cause close to your heart, knowing your "why" will help you navigate life's challenges with clarity and determination.

2. **Vision Beyond the Present: Seeing the Unseen**

   Linda's mother didn't know exactly what was out there beyond Judson Street, but she knew it had to be better. This vision wasn't about knowing all the answers; it was about having faith that something better existed and the determination to seek it out. Vision is about seeing beyond your current circumstances and imagining a future that is better, brighter, and more aligned with your purpose. It's the ability to look past what is and see what could be.

   - **Reflection:** What is your vision for your life? Can you close your eyes and see the future you want to create? Your vision doesn't have to be fully formed or perfect, but it needs to exist as a guiding star. It's what will shape your decisions, inspire your actions, and keep you focused on your goals, even when the path forward is unclear.

   - **Strategy:** Develop your vision by imagining the life you want to lead, the impact you want to have, and the legacy you want to leave. Write it down, create a vision board, or talk about it with someone you trust. Revisit this vision regularly and allow it to evolve as you grow. Your vision will serve as both a compass and a source of motivation, helping you to stay aligned with your purpose and navigate life's twists and turns with confidence.

3. **Taking Responsibility and Leading: Stepping Into the Arena**

   One of the most powerful aspects of Linda's journey was her decision to step up when no one else would. When she couldn't find a leader to take on the challenge of transforming Strawberry Mansion High

School, she looked in the mirror and realized that she was the one who needed to lead. This decision wasn't made lightly—it came with fear, resistance, and significant risk. But it also came with a deep sense of responsibility, rooted in her commitment to her students and her belief that they deserved better.

- **Reflection:** Are there areas in your life where you are waiting for someone else to lead? What would happen if you stepped up instead? Leadership isn't about having all the answers or being without fear; it's about recognizing when you are the one who must take action, even when it's uncomfortable or frightening. Reflect on the moments in your life when you've chosen to lead—what drove you to step up, and what impact did it have?

- **Strategy:** Leadership begins with the decision to take responsibility for the change you want to see. Identify an area in your life—whether personal, professional, or within your community—where your leadership is needed. Take small, deliberate steps to step into this role, knowing that true leadership often involves stepping into the unknown with courage and conviction. Remember, leadership is not about waiting for perfect conditions; it's about making the decision to lead, even in the face of uncertainty.

4. **The Power of Incremental Change: Building a Legacy Step by Step**

Linda's approach to transforming her school wasn't about making grand gestures or expecting overnight success. Instead, it was about implementing small, consistent changes that, over time, led to significant transformation. She focused on creating a safe and nurturing environment where students could thrive, knowing that these small changes would eventually lead to larger, more lasting results. This approach underscores the power of incremental change—how small, deliberate actions can build a foundation for significant impact.

- **Reflection:** What are your long-term goals? How can you break them down into small, manageable steps? Too often, we get overwhelmed by the size of our goals, leading to inaction or burnout. Reflect on the small, consistent actions you can take today that will move you closer to your desired outcome. What systems can you put in place to ensure that these actions are sustained over time?

- **Strategy:** Focus on the process rather than the outcome. Identify one small change you can make today that aligns with your long-term goals. Commit to this change and track your progress over time. As you build momentum, you'll find that these small steps lead to significant growth and transformation. Remember, meaningful change doesn't always happen overnight, but with persistence and patience, your efforts will accumulate into lasting impact.

5. **Leading with Love: The Heart of True Leadership**

Perhaps the most profound lesson from Linda's story is the power of leading with love. Every day, Linda ended her announcements with a simple yet powerful message to her students: "And remember, if nobody told you they loved you today, you remember I do, and I always will." This wasn't just a platitude—it was a commitment, a declaration of her unwavering belief in her students' worth. Leading with love means putting the well-being of others at the forefront of your actions. It means creating an environment where people feel valued, supported, and empowered to reach their full potential.

- **Reflection:** Are you leading with love in your relationships, whether at home, at work, or in your community? How do you show others that you genuinely care about them? Reflect on the ways in which you express love and compassion in your leadership.

What impact does this have on those around you, and how might you deepen these expressions of care and support?

- **Strategy:** Make a conscious effort to lead with love in all areas of your life. This might involve simple acts of kindness, words of encouragement, or simply being present and attentive to the needs of others. Build trust through consistent, loving actions, and create a space where people feel safe, valued, and supported. Remember, love is the foundation of trust, and trust is the foundation of all meaningful relationships and change. By leading with love, you not only inspire others—you also cultivate an environment where true transformation can take place.

Linda Cliatt-Wayman's story is a powerful reminder that each of us has the capacity to make a difference. We may not all be called to transform a struggling school, but we are all called to lead in some way. Whether it's in our families, our workplaces, or our communities, we have the power to create change when we lead with purpose, vision, responsibility, and love.

As you reflect on Linda's journey, consider this: Who in your life has been that light for you? Who has shown you that there is more to life than the circumstances you find yourself in? And more importantly, who are you being that light for? Linda's story challenges all of us to look beyond our own lives and ask ourselves, "So what, now what?" How can we take the challenges we face and turn them into opportunities for growth, for change, for making a difference in the lives of others? Because, as Linda's story so powerfully illustrates, when we choose to lead with love, when we choose to care deeply and act boldly, we can change the world—one life at a time.

As you move forward in your own journey, take these lessons to heart. Define your purpose, create a vision for your life, step up to lead when needed, embrace the power of incremental change, and always, always lead with love. In doing so, you'll not only transform your own life—you'll also be a light for others, helping them to see that there is more to life than the streets they walk every day.

To hear Linda's story and learn more about her journey:

https://drivingchangepodcast.com/podcast/linda-cliatt-wayman/

https://principalwayman.com/

CHAPTER 11

Craig Lucas

From Despair to Destiny: The Power of Perseverance and Service to Others

Craig Lucas' story isn't just about professional highs and personal lows; it's about the emotional journey of a man who learned, through tremendous trial and loss, what his true purpose was. It's about resilience, vulnerability, and, ultimately, the power of transformation. As I reflect on Craig's life, it becomes clear that his story speaks to something deep within us all—a longing to overcome, redefine ourselves, and give more of who we are to the world.

Craig grew up in Strongsville, Ohio, where life seemed simple, yet it held the seeds of something much bigger. As a kid, Craig was a boy with endless curiosity. He wasn't the kind of kid who fit neatly into one box. One moment he could be thriving on the kickball field and the next, he might be down by the creek bed, searching for fossils with friends. He lived on the edges, never settling into one identity, always exploring different sides of himself. In those moments, whether competing in sports or wandering in nature, a quiet determination was beginning to take root. Craig wasn't the center of any one thing, but he was connected to everything around him—an awareness of people, moments, and possibilities.

But what really stands out about Craig's childhood is the influence of his mother. A sixth-grade English teacher for 36 years, she was the one who instilled in Craig a sense of discipline and attention to detail. Her commitment to her students, her belief in doing things with care, left a lasting imprint on Craig's soul. She wasn't just a teacher in the classroom; she was a teacher in life. Her lessons, though simple, carried weight—how to speak with intention, how to care about the people you're serving, and how to always go the extra mile.

One memory stands out. Each year, she would take her students to a local upscale restaurant to teach them etiquette—a lesson far beyond academics. She was showing them how to navigate the world with confidence, to see themselves as worthy of more than their current circumstances. It was a gesture of love and with that, Craig absorbed something profound. He learned that details and intention matter, not just because they're tasks to be completed, but because they reflect how much we care about the people in front of us.

Craig's mother's influence stretched far beyond the classroom, and it's no surprise that Craig took those lessons with him into his own life. But like most all of us, his journey wasn't without struggle. After a childhood spent working odd jobs—from shoveling snow to selling vegetables from a red wagon—Craig graduated from college and found his way into the world of investment advisory. And this is where his story begins to take shape.

Craig entered the financial world with ambition, determination, and a work ethic that set him apart. He started his career on Valentine's Day 1987, just months before the stock market crash that would send shockwaves through the industry. For most, the crash was a devastating blow, but for Craig, it was a moment of clarity. He watched as seasoned advisors hesitated, unsure of how to navigate the crisis. Clients' phones were ringing off the hook, but no one was answering. Craig made a decision that would shape his future—he would answer the calls. He would show up for his clients when they needed him most, no matter how hard it was.

That choice, that simple act of showing up, became the cornerstone of Craig's approach to life and business. He wasn't just an advisor; he was someone his clients could rely on in their darkest moments. He built a billion-dollar practice on that foundation—not by being flashy or cutting corners, but by asking the tough questions, listening, and caring deeply about the people he served.

But life has a way of testing even the strongest foundations. In 2005, Craig's world came crashing down in a way he never saw coming. A required form unknowingly not executed that called into question how he was supporting his brother-in-law, led to a compliance investigation that spiraled out of control. In a matter of days, everything he had built over two decades was gone—his job, his financial stability, his identity. He was fired, and the future he had worked so hard to build evaporated overnight.

Craig describes that period as one of complete darkness. It wasn't just the loss of a job—it was the loss of everything he thought he was. Imagine for a moment what that feels like. To wake up one day and realize that the life you've built, the reputation you've earned, has been shattered. He was no longer Craig Lucas, the successful financial advisor. He was Craig Lucas, the man who had lost everything.

It was a time of deep despair. Craig had two young children, a wife, and a mountain of uncertainty before him. He felt like a blind man in a dark room, desperately trying to find his way. But in the midst of that darkness, Craig made a choice—he wasn't going to give up. He wasn't going to let this moment define him. There were days when the fear was overwhelming, when the uncertainty felt too heavy to bear. But Craig leaned into the one thing he knew he could control—his response.

It wasn't easy. There were countless sleepless nights, hours spent trying to piece together a future that felt distant and uncertain. But Craig held onto the belief that his story wasn't over. That the best was yet to come, even if he couldn't see it yet. He clung to hope, not just for himself, but for his family. And in those moments of darkness, something beautiful began to emerge—an understanding that his purpose was bigger than his career, bigger than his success. His purpose was rooted in something deeper, something unshakable.

Craig did not return to the financial world. Instead, he took time to reflect, heal, and rebuild—not just his career, but his sense of self. In 2014, working In the pre-dawn and weekend hours during a work chapter where Craig became relevant in an all-together different industry, he founded Authenica, a company that would reflect the values he had come to hold dear—vulnerability, connection, and transformation. Authenica wasn't just about creating corporate events; it was about creating experiences that pushed people to confront their own stories, fears, and truths.

The magic of Authenica lies not in its venues or its guest lists, though both are impressive. It lies in Craig's ability to create spaces where people feel safe enough to be vulnerable. It's in the way he asks the right questions, the tough questions, the ones that force you to look inward and find answers you didn't know you were searching for. Craig's events are about more than networking or team building—they're about transformation. They're about helping people connect with their true selves, just as Craig had to connect with his during his darkest days.

Craig's story is a testament to the power of resilience, of trusting the process even when the path ahead is unclear. It's a reminder that we are not defined by our failures, but by how we respond to them. Craig didn't let his darkest moment be the end of his story. He used it as the foundation for something new, something powerful, something that would touch the lives of others in ways he couldn't have imagined.

Today, Craig stands as a beacon of hope for those who feel lost, for those who wonder if they'll ever find their way out of the darkness. His story is a reminder that even in our lowest moments, there is always the possibility of redemption, of reinvention. Craig's journey isn't just about bouncing back—it's about becoming more of who he was always meant to be.

And that's the beauty of Craig's story. It's not about the fall; it's about the rise. It's about finding purpose in the pain, strength in the struggle, and hope in the hardship. It's about realizing that sometimes, when everything falls apart, it's because something even greater is waiting to be built. Craig's life is proof that no matter how dark the night, the dawn always comes. And when it does, it brings with it the possibility of a new beginning—a beginning that can be even more beautiful than what came before.

Reflection: Embracing the Journey Through Loss, Resilience, and Purpose

Craig Lucas's story reminds us that life's greatest lessons are often learned in the midst of adversity. His journey wasn't just about overcoming the loss of a successful career; it was about rediscovering himself, redefining success, and using his experience to serve others. As we reflect on Craig's story, there are five key lessons that can help each of us navigate our own challenges and build a life of deeper purpose. These lessons, paired with actionable strategies, can guide us toward resilience, authenticity, and meaningful connection.

1. **Redefining Success: It's About Who You Become, Not What You Achieve**

   Craig's story shows us that true success isn't just about external achievements but about the person we become along the way. After losing his business and financial security, Craig discovered that success is more about character, growth, and impact on others than wealth or recognition.

   - **Reflection:** For much of Craig's life, success was defined by external markers—a billion-dollar book of business, financial security, and professional recognition. But when those things were taken away, Craig was forced to ask himself a deeper question: What does success truly mean? His answer came not in the form of rebuilding his financial empire, but in the discovery that success is about who you become, not what you achieve. True success lies in character, growth, and the impact we have on others.
   - **Strategy:** Take a step back from your current goals and achievements. Instead of focusing solely on what you're working toward, reflect on who you are becoming in the process. Ask yourself: What qualities do I want to embody as I pursue my goals? How can I redefine success to include personal growth,

resilience, and deeper connection with others? Write down three ways you can shift your focus from external achievements to internal growth this week.

2. **Leaning Into the Darkness: Transforming Struggles into Growth**

When Craig lost everything, he made a conscious choice to lean into his challenges and trust that they could lead to growth. This decision reminds us that hardship, while painful, can be a powerful catalyst for transformation.

- **Reflection:** When Craig lost everything, it would have been easy for him to give in to bitterness, anger, or despair. Instead, he made a conscious choice to lean into the darkness and trust that his struggles would lead to something greater. He didn't know what was next, but he knew the challenges he faced held the potential for transformation. In life, it's often the moments of pain and hardship that push us to grow in ways we never imagined.

- **Strategy:** Identify a current struggle or challenge you are facing. Instead of avoiding it or letting it overwhelm you, consider how this difficulty might be an opportunity for growth. What lessons can you learn from it? What strengths are you developing as a result? Journal about how this struggle could shape you into a stronger, wiser, or more compassionate person, and identify one small action you can take today to embrace the challenge rather than resist it.

3. **Letting Go of Control: Finding Peace in the Present Moment**

One of Craig's most powerful lessons was recognizing the futility of trying to control everything. By focusing on the present and trusting the process, he found peace amidst his loss. This approach teaches us that true power lies in knowing what we can control and releasing the rest.

- **Reflection:** One of the most powerful lessons Craig learned was that he couldn't control everything. When his career crumbled, he realized that trying to manage every aspect of the situation would only lead to more overwhelm. Instead, he focused on what was directly in front of him—being present with his family, taking one step at a time, and trusting the process. Letting go of control doesn't mean giving up; it means recognizing where your power lies and releasing the need to manage everything.

- **Strategy:** Reflect on an area of your life where you feel the need to control every detail—whether it's your career, relationships, or personal goals. Consider what aspects of the situation are truly within your control, and what you need to release. Practice letting go by focusing on the present moment. Take three deep breaths and write down one thing you can control today (your attitude, your effort, your responses), and one thing you will consciously release.

4. **Vulnerability as Strength: Building Authentic Connections**

   Craig's journey taught him the strength in vulnerability. By sharing his struggles, he was able to forge deeper connections with those around him. True connection isn't built on perfection but on authenticity.

   - **Reflection:** Throughout his journey, Craig discovered the power of vulnerability. Instead of hiding his pain or pretending everything was fine, he chose to share his struggles openly with those around him. This openness didn't make him weak—it made him stronger. It allowed him to build deeper, more meaningful relationships. Vulnerability isn't about exposing every detail of our lives; it's about allowing others to see us as we truly are, flaws and all. True connection is built on authenticity.

   - **Strategy:** Reflect on a relationship in your life where you may be holding back or hiding parts of yourself. Consider how you can open up in a way that fosters trust and connection. This week, choose one person with whom you will share something you've been struggling with—whether it's a fear, a challenge, or an uncertainty. Notice how being vulnerable changes the dynamic of the relationship and deepens your connection.

5. **Purpose Through Pain: Using Your Story to Serve Others**

Craig found purpose in his pain by helping others through Authenica, using his story to inspire others to find meaning in their struggles. This reminds us that our deepest purpose often arises from our greatest challenges.

- **Reflection:** After losing his career, Craig could have focused on rebuilding his own wealth or reputation. Instead, he chose to use his story to serve others. Through Authenica, Craig now helps people connect with their true potential, using his own experiences to inspire others to find meaning in their struggles. The truth is, our deepest purpose often emerges from our greatest pain. The challenges we overcome give us the empathy and wisdom to help others on their journey.

- **Strategy:** Think about a difficult experience you've been through. How has that experience shaped you? How has it given you insight, strength, or empathy? Now, consider how you can use that experience to serve others. Is there someone in your life who is going through something similar? How can you offer support, encouragement, or guidance? Make a commitment to reach out to someone this week and share your story in a way that offers hope or insight.

Craig Lucas's story is one of resilience, authenticity, and purpose. The lessons he learned through his journey are not just for him—they are for all of us. Life will always have its challenges, but the way we respond to those challenges is what shapes our character, our relationships, and our legacy.

As you reflect on Craig's story and the five lessons we've explored, I encourage you to take intentional steps toward applying these strategies in your own life. Whether it's redefining success, embracing vulnerability, or using your pain to serve others, know that your journey—like Craig's—is filled with opportunities for growth, connection, and purpose.

Your story, with all its ups and downs, is unfolding. And just like Craig, you have the power to turn your struggles into strength, your challenges into opportunities, and your life into a testament to the resilience of the human spirit.

To learn more about Craig's work and his journey:

https://authenica.com/

https://drivingchangepodcast.com/podcast/navigating-adversity-finding-purpose-craig-lucas/

CHAPTER 12

Tricia Downing

The Road to Resilience: Transforming Tragedy into Triumph

Imagine, if you will, the moment when everything you've built your life around is ripped away in an instant. It's the kind of event that forces you to confront the deepest parts of yourself, to discover what you're truly made of. This is the story of Tricia Downing, a woman whose life was forever altered in the blink of an eye—a story that speaks not just to the power of the human spirit, but to the resilience that lies within each of us, waiting to be awakened.

Tricia's life was defined by one word: athlete. From an early age, her world revolved around sports. The water was her first love; she spent her childhood swimming, a passion that soon evolved into diving and gymnastics. But it was cycling that truly captured her heart. Cycling was freedom, it was exhilaration, it was the path she had chosen to pursue with all her being. Tricia wasn't just participating—she was excelling, pushing herself to the limits, chasing the dream of becoming a professional cyclist.

In the summer of 2000, Tricia was on fire. She had just returned from a grueling but rewarding cross-country racing tour, more certain than ever that cycling was her destiny. This was more than a hobby; it was the very essence of who she was. Every pedal stroke, every hill climb, every race was another step closer to her dream. She was strong, determined, and unstoppable—or so it seemed.

But life, as we all know, has a way of challenging us in ways we never expect. On a sunny September day, during what should have been a routine training ride with a friend, Tricia's life was irrevocably changed. They were heading back into Denver after a climb up Lookout Mountain, a favorite spot for local cyclists. Tricia was following closely behind her friend Matt when they approached an intersection. A car was coming from the opposite direction, preparing to make a left turn.

In an instant, everything slowed down. Tricia watched as Matt narrowly avoided a collision, swerving around the car. Relief washed over her—he was safe. But then she realized, with heart-stopping clarity, that the driver hadn't seen her. There was no time to react, no time to stop. Tricia collided with the front of the car, her body flipping up and crashing into the windshield before she was thrown to the ground.

In that moment, as she lay on the pavement, something inside her knew that life as she had known it was over. She couldn't feel her legs. They were there, but they felt like they were floating above her, detached from her body. The world around her seemed to fade as she was rushed to the hospital, where a battery of tests confirmed her worst fear—she was paralyzed from the chest down. The doctors told her family she would likely never walk again.

Tricia was 31 years old, in the prime of her life, and suddenly everything she had worked for, everything she had dreamed of, seemed to be lost. How do you even begin to process something like that? How do you find the strength to move forward when the very thing that defined you has been taken away?

The early days after the accident were a blur of pain and disbelief. Tricia's injuries were severe—broken ribs, a fractured neck, and multiple fractured vertebrae that had damaged her spinal cord. The pain was excruciating, a constant reminder of what she had lost. She spent weeks in the ICU, her body fighting to stabilize after surgery to place rods in her back. Even after she was transferred to Craig Hospital, one of the top spinal cord injury rehabilitation centers in the country, the battle was far from over.

It wasn't just about learning to live in a wheelchair—it was about surviving. Tricia was still incredibly sick, her body struggling to heal. One of the most difficult challenges she faced was a pressure sore on her back, a wound that refused to heal and eventually required surgery. For six weeks, she had to lie flat on her back, unable to sit up, unable to do anything but lie there and wait. The physical pain was matched only by the emotional agony of feeling trapped, like a caged animal with no way out.

This was rock bottom. Tricia was alone with her thoughts, with the realization that her life had been shattered. The future she had envisioned, the life she had planned, was gone. The dreams of becoming a professional cyclist, of conquering the world on two wheels, had been replaced by a harsh new reality—one where even the simplest tasks, like getting out of bed or driving a car, required a monumental effort.

But here's the thing about Tricia Downing—she is a fighter. Even in those darkest moments, when it seemed like there was no light at the end of the tunnel, something deep within her refused to give up. Maybe it was the athlete in her, the part of her that had always thrived on challenges, on pushing herself to the limit. Or maybe it was the support of her family, who stood by her side, encouraging her to keep going. Or perhaps it was the memory of the disabled athletes she had once ridden with as a tandem pilot, athletes who had shown her that life didn't have to stop just because you faced obstacles.

Whatever it was, it sparked a fire within her. Tricia decided that if she couldn't walk, if she couldn't ride a bike the way she used to, she would find another way to be an athlete. She wasn't going to let paralysis define her. She wasn't going to let it be the end of her story.

The road to recovery was long and arduous. There were days when Tricia couldn't see beyond the pain, days when the enormity of what she had lost threatened to overwhelm her. But she kept moving forward, one small step at a time. She learned to navigate the world in a wheelchair, to build up her upper body strength so she could transfer herself from her chair to a bed, to a car, to whatever life required. She learned to drive with her hands, to live independently in a world that wasn't designed for people with disabilities.

And slowly, but surely, she began to rebuild her life. It wasn't easy. Everywhere she went, she was confronted with reminders of what she could no longer do. People looked at her differently, treated her differently, as if she were somehow less because she was in a wheelchair. But Tricia refused to let that get to her. She knew that she was still the same person, still the same athlete, even if her body had changed.

It was during this time that Tricia made a pivotal decision—she wasn't just going to survive, she was going to thrive. She began to explore new possibilities, new ways to challenge herself. She discovered hand cycling, a sport that allowed her to experience the thrill of cycling again, even if it was in a different form. She started swimming again, even though her first attempts left her feeling defeated and frustrated. And she found triathlons—a new arena where she could push her limits, where she could compete not just against others, but against herself.

The journey wasn't without setbacks. Tricia faced more surgeries, more pain, more moments of doubt. But with each challenge, she grew stronger. She learned to redefine success, to measure it not by what she could no longer do, but by what she was still capable of. And what she was capable of was nothing short of extraordinary.

Tricia became the first female paraplegic to complete an Ironman triathlon, a grueling test of endurance that includes a 2.4-mile swim, a 112-mile bike ride, and a 26.2-mile marathon. She competed in over a hundred races, pushing herself further and harder than she ever thought possible. And she didn't stop there. She set her sights on the Paralympics, representing Team USA in shooting at the 2016 Games in Rio de Janeiro.

Tricia's story is not just one of physical triumph—it's a story of mental and emotional resilience, of the power of the human spirit to overcome even the most devastating of circumstances. It's about finding hope in darkest of days, about refusing to let life's challenges define you. It's about the strength that comes from within, the strength that we all have, even if we don't always realize it.

As we look at Tricia's journey, we are reminded that life is full of unexpected twists and turns. We can't always control what happens to us, but we can control how we respond. Tricia chose to respond with courage, with determination, with a refusal to let her circumstances dictate her future. And in doing so, she has become a beacon of hope, an inspiration to anyone who has ever faced a challenge that seemed insurmountable.

Tricia Downing's story is a testament to the power of the human spirit, to the resilience that lies within each of us, waiting to be awakened. It's a reminder that no matter what life throws our way, we have the strength to overcome it. We have the power to redefine our lives, to find new paths to success and fulfillment, to keep moving forward, even when the road ahead seems impossible.

In the end, Tricia's journey is not just about overcoming paralysis—it's about discovering the true meaning of strength, of resilience, and of hope. It's about showing us all that, no matter what challenges we face, we have the power to rise above them, to create a new future for ourselves, one step at a time.

In her memoir, *Cycle of Hope: A Journey from Paralysis to Possibility*, Tricia shares not just the triumphs, but the struggles, the moments of doubt, and the hard-won lessons she learned along the way. She writes about the importance of setting goals, no matter how small, and of celebrating each step forward, even when progress feels slow. She emphasizes the value of hope and the belief that the future can be better than the present—if you're willing to put in the work to make it so.

Reflection: Lessons from Tricia's Journey

As we reflect on Tricia Downing's extraordinary journey, we find ourselves confronted with powerful lessons that resonate deeply within each of us. Her story is not just about overcoming physical challenges; it's about the resilience of the human spirit, the power of hope, and the courage to redefine one's life in the face of adversity. Let's take a moment to explore five key life lessons that emerge from Tricia's experience—lessons that can guide us on our own journeys, no matter the challenges we face.

1. **Redefine Success on Your Own Terms**

   Tricia's story teaches us the importance of redefining success, especially when life takes an unexpected turn. When she lost the ability to walk, Tricia didn't lose her identity as an athlete—she redefined what it meant to be one. Instead of focusing on what she could no longer do, she found new ways to challenge herself, ultimately becoming a world-class athlete in a different realm.

   - **Reflection:** How do you define success in your own life? Is it tied to specific outcomes or external achievements? What would it look like to redefine success based on your current circumstances? Remember, success isn't about achieving a specific goal—it's about the journey, the growth, and the resilience you demonstrate along the way.

   - **Strategy:** Take some time to reflect on your current goals. Are they still aligned with who you are and where you are in life? If not, consider redefining them in a way that honors your current reality while still challenging you to grow and evolve.

2. **Embrace the Power of Incremental Progress**

   Tricia's recovery and return to athletics were not the result of a single, monumental effort, but of countless small steps taken consistently over time. She understood that big goals are achieved through incremental progress, and she focused on what she could do each day to move closer to her goals.

   - **Reflection:** Are you overwhelmed by the size of the challenges you face? Do you find yourself paralyzed by the distance between where you are and where you want to be? Tricia's journey reminds us that we don't have to conquer the entire mountain in a single day—sometimes, it's enough to take just one step.

- **Strategy:** Break down your goals into smaller, manageable tasks. Celebrate each small victory, and use it as motivation to keep moving forward. Remember, progress is progress, no matter how small it may seem.

3. **Find Strength in Your Support System**

Throughout her journey, Tricia drew strength from the people around her—her family, friends, and the community of athletes with disabilities who inspired her. She didn't try to go it alone; instead, she leaned on those who cared for her, allowing their support to bolster her own resilience.

- **Reflection:** Who are the people in your life that you can lean on during tough times? Are you allowing yourself to be supported, or are you trying to carry the burden alone? Remember, it's okay to ask for help. In fact, it's a sign of strength, not weakness.

- **Strategy:** Identify your support system—those people who uplift you and offer encouragement. Don't be afraid to reach out to them when you need support. At the same time, consider how you can be a source of strength for others in your life.

4. **Embrace the Power of Perspective**

Tricia's ability to shift her focus from what she had lost to what she could still achieve was a pivotal moment in her journey. By changing her perspective, she transformed her situation from one of despair to one of possibility. She recognized that while she couldn't change her circumstances, she could change how she responded to them.

- **Reflection:** How do you perceive the challenges you face? Are they insurmountable obstacles, or opportunities for growth and transformation? Your perspective can either limit you or empower you—Tricia's story is a powerful reminder of the latter.

- **Strategy:** Practice reframing your challenges. Instead of viewing them as setbacks, consider what opportunities they might present. What can you learn from them? How can they help you grow? By changing your perspective, you can change your entire experience.

5. **Cultivate Resilience Through Hope and Action**

   Tricia's journey was fueled by an unwavering sense of hope—a belief that her future could be better, and that she had a role in making it so. But hope alone wasn't enough; it was Tricia's willingness to take action, even on the hardest days, that made all the difference. She didn't let fear of failure hold her back—instead, she used it as motivation to keep pushing forward.

   - **Reflection:** What role does hope play in your life? Do you allow yourself to dream of a better future, or are you afraid to hope for fear of disappointment? Tricia's story reminds us that hope is a powerful force, but it must be coupled with action.

   - **Strategy:** Take a step, no matter how small, toward the future you hope for. Whether it's setting a new goal, seeking out support, or simply getting out of bed on a tough day, every action you take moves you closer to the life you want. Remember, resilience is built one step at a time.

Tricia's story is one of extraordinary courage and resilience, but it's also a story that each of us can learn from. We all face challenges—some visible, some hidden—but it's how we respond to those challenges that defines us. Take these lessons to heart and use them to fuel your own journey of growth, healing, and transformation. Remember, you have within you the strength to overcome, the resilience to keep going, and the power to redefine your life in ways you never thought possible.

To hear Tricia's story and learn more about her journey:

https://drivingchangepodcast.com/podcast/tricia-downingcycle-of-hope/

https://triciadowning.com/

https://www.amazon.com/Cycle-Hope-Journey-Paralysis-Possibility-ebook/dp/B006J8M70S

CONCLUSION

The Power of *Your* Story

As I sit here reflecting on the incredible stories we've journeyed through together in this book, I'm struck by a profound truth that has been the foundation of my life's work: every story matters. Each one of the lives we've explored—Chad's determination in the face of darkness, Dawn's strength through trials, Hilary's battle with inner voices, Craig's relentless pursuit of purpose, James's unwavering faith, Johnny's journey through grief, Katie's spirit of survival, Linda's quiet resilience, Tricia's indomitable will to overcome, Andy's unexpected courage, Ben's quest for redemption, and Jordan's search for identity—holds a mirror to the deepest parts of our own souls. These stories aren't just about them; they're about you, me, and every person who has ever wondered if their life has significance.

What makes these stories so powerful is not just the obstacles they faced, but the decisions they made in the face of those obstacles. Each of these individuals had a choice: to surrender to their circumstances or to rise above them, to let the weight of their challenges crush them or to use those very challenges as fuel to propel them forward. In the same way, you have that choice. Your story has been unfolding your entire life, and with every new chapter comes a decision: how will you respond? Will you let the setbacks define you, or will you turn them into steppingstones? Will you write a narrative of defeat, or will you craft a story of resilience and triumph?

I've spent years studying the human brain, understanding how we're wired, how we think, and how we connect. And if there's one thing I've learned, it's that our stories are our most powerful tools. They shape us, define us, and ultimately, they guide us. But let me tell you something even more important: your story—yes, your story—is not only powerful, it's essential.

Each of the people whose stories we've explored in this book faced unimaginable challenges. Some were born into them, others stumbled into them unexpectedly. But every single one of them made a decision, at some point, to see their life not as a series of random, disconnected events, but as a meaningful journey. They chose to find purpose in their pain, strength in their struggle, and hope in their darkest hours.

Think about that for a moment. What if Chad had seen his blindness as the end of his journey? What if Dawn had given up when the storms of life grew too strong? What if Hilary had let those inner voices of doubt drown out her will to fight? And what about you? What if you stopped seeing your challenges as obstacles and started seeing them as opportunities? Opportunities to grow, to learn, to change, and to inspire others with your resilience.

Purpose isn't something that just happens to us. It's something we uncover, something we choose to see in our lives. It's found in the everyday moments, in the trials and triumphs, in the connections we make with others. And it's most powerfully found when we take the time to understand and embrace our own stories.

I want you to take a moment right now and think about the moments in your life that have shaped you. The moments that have tested you, pushed you to your limits, and maybe even broken you. Now, ask yourself: what did I learn from those moments? How did they shape the person I've become today? Often, we move through life so quickly that we don't stop to reflect on the impact of our experiences. We don't take the time to connect the dots, to see how everything we've been through has brought us to this very moment. But when you do stop and reflect, when you do connect those dots, a new story begins to emerge—a story of strength, resilience, and purpose.

These are not just stories to be admired—they are stories to be learned from. Because within each of them is a lesson, a strategy, a glimpse of what's possible when we choose to embrace our own story with the same courage, the same determination, the same hope.

So, what about you?

What is your story telling you? Are you listening to the voices that say you're not enough, that your past defines you, that your failures are final? Or are you ready to rewrite that narrative? Are you ready to see your life as a journey filled with purpose, potential, and possibility?

One of the most powerful things I've learned in my life is that our brains are wired for connection. We are wired to connect with others through stories. But more importantly, we are wired to connect with ourselves through our own story. When we understand our story—when we truly take the time to see it, to feel it, to own it—we unlock a power within us that is unstoppable.

You see, it's easy to look at the stories of others and think, "That's amazing, but I could never do that." But here's the truth: you are already doing it. Every day, you are writing the story of your life. Every decision, every action, every choice you make is a part of your narrative. The question isn't whether you have a story worth telling—the question is whether you're ready to own it, to embrace it, and to share it with the world.

Your story has brought you to this moment. Every joy, every heartbreak, every victory, every defeat—it's all part of the story that has shaped you into who you are today. But here's the thing: your story isn't finished. The next chapter is waiting to be written, and you are the author.

There is power in that realization. There is freedom in knowing that no matter what has happened in the past, no matter what mistakes you've made, no matter what challenges you've faced, you have the ability to write a new chapter. You can take control of your story and steer it in a new direction.

Think about the people whose stories we've explored in this book. Chad didn't let his blindness define him; he found a new vision, one that was more powerful than physical sight. Dawn didn't let the storms of life defeat her; she discovered a strength within herself that was unshakable. Hilary didn't let the voices of doubt control her; she found her own voice and used it to rewrite her story. Each of these individuals faced incredible challenges, but they didn't let those challenges define them. They chose to take control of their stories, to rewrite their narratives, and to create lives filled with purpose, strength, and hope.

And that's what I want for you. I want you to see the power in your own story. I want you to recognize that your past doesn't define you—your response to it does. I want you to understand that the challenges you've faced, the struggles you've endured, and the victories you've won are all part of a larger narrative, one that is filled with purpose and potential.

Whatever your story, know this: it matters. It matters because it's yours. It matters because it's the foundation of your purpose. And it matters because the world needs to hear it.

I want to leave you with this: the most important thing you can do from this moment on is to own your story. Embrace it fully. See the beauty in the brokenness, the strength in the struggle, and the hope in the hardship. Because when you do, you will not only change your life—you will change the lives of those around you.

Your story is your power. Your purpose is your gift. And the world is waiting for you to step into both.

So, as you close this book, I challenge you to take a moment to reflect on your own journey. What have you learned? How have you grown? Where are you going next?

Remember, your story is not over. It's just beginning. And the best is yet to come.

Go out there and write the next chapter of your life with courage, with purpose, and with the unwavering belief that your story matters. Because it does. More than you may ever know.

Thank you for joining me on this journey. I can't wait to see the incredible stories you will create and the impact you will make in the world.

—Jeff

Made in the USA
Monee, IL
08 November 2026

67961521R00080